Equity in Discourse for Mathematics Education

Mathematics Education Library
VOLUME 55

For further volumes:
http://www.springer.com/series/6276

Beth Herbel-Eisenmann • Jeffrey Choppin
David Wagner • David Pimm
Editors

Equity in Discourse for Mathematics Education

Theories, Practices, and Policies

Springer

Editors
Beth Herbel-Eisenmann
Michigan State University
329 Erickson Hall
East Lansing, MI 48824
USA
bhe@msu.edu

Jeffrey Choppin
Warner School of Education
University of Rochester
Box 270425
Rochester, NY 14627
USA
jchoppin@warner.rochester.edu

David Wagner
Faculty of Education
University of New Brunswick
P.O. Box 4400
Fredericton, NB E3B 5A3
Canada
dwagner@unb.ca

David Pimm
Faculty of Education
Simon Fraser University
Burnaby, BC V5A 1S6
Canada
david.pimm@ualberta.ca

ISBN 978-94-007-9365-1 ISBN 978-94-007-2813-4 (eBook)
DOI 10.1007/978-94-007-2813-4
Springer Dordrecht Heidelberg London New York

Printed on acid-free paper

Springer is part of Springer Science+Business Media (www.springer.com)

Foreword

Discourse and Equity: The Simultaneous Challenge of Epistemological and Social Access

School mathematics remains a powerful social filter, and understanding and explaining access to and success in school mathematics has been of considerable interest to the research community for some time now. Captured as the 'social turn', Lerman (2000) examined research in mathematics education that took as its starting point a socially situated mathematics classroom – a classroom located in a wider social structure, one simultaneously constitutive of and constituted by social relations within it. The social turn reflected recognition that considerable progress in understanding mathematical learning was largely achieved by means of a focus on cognition, backgrounding the social distribution of success and failure. A broader set of theoretical resources and methodological tools was required and sociocultural theories were increasingly drawn on and drawn in. Contained within this turn, but not reducible to it, was a concern with equity and the power of school mathematics both to enable and to exclude.

Alongside a concern with the social was increasing interest in what can now be captured as the 'discursive turn'; an appreciation for discourse as constitutive of participation in school mathematics classrooms. Mathematics classroom communication attracted increasing attention, bringing into focus discourse and social interaction. Yet, whilst inextricably related, issues of equity were taken up largely within the social turn while issues of mathematics classroom communication travelled a parallel path. A decade ago, as I worked on a manuscript on teaching and learning mathematics in multilingual classrooms (Adler 2001), these parallel paths were highly visible. I noted "a continuing disjuncture between research on communication in bi-/multilingual mathematics classrooms on the one hand, and what could be described as more mainstream research on communication in *the* mathematics classroom on the other" (p. 13; *italics in original*). Insights into pedagogic communication and, particularly, into teacher mediation in mathematics classrooms, foregrounded in research in the multilingual classroom context, were largely 'ignored' in mainstream research on classroom communication.

There has since been significant progress in all these areas of research in mathematics education, particularly in unpacking participation in mathematical discourse in school classrooms. School mathematics is a pedagogised mathematics (Bernstein 2000; Singh 2002), shot through with assumptions about learners and learning, teachers and teaching, requiring an understanding of how macro and micro social forces are simultaneously at work in pedagogic practice. For Bernstein, what is constituted as mathematics in pedagogic discourse, how this is constituted and what this means for 'acquisition', in his terms, cannot be taken for granted. Simultaneous attention to and, through this, more nuanced and critical understanding of who participates in which kinds of pedagogic practice, and with what consequences for mathematics learning, remains a significant challenge, a challenge embraced in this book.

In post-apartheid South Africa, increasing the quality, quantity and social distribution of participation in education at all levels is an on-going concern for research, policy and practice. In higher education research, useful distinctions have been made between *institutional* and *epistemological* access (Morrow 1993), and, more recently, social access (Cross et al. 2010). There is evidence that whilst universities have opened up institutional access (who gets to enter the university for degree study), as well as developed policies in support of both epistemological and social access, the practical implementation of appropriate pedagogies on the one hand and strategies for dealing with varying levels and forms of cultural capital in the social spaces within the institution on the other has proved far more intractable. While the South African context is distinct, the illumination of the interweaving of institutional, epistemological and social access to educational opportunity lights up the challenges of discourse and equity as discussed in this book. As Guitérrez describes in Chap. 2, these challenges are about access and achievement on the one hand *and* about identity and power on the other.

David Wagner, Beth Herbel-Eisenmann and Jeffrey Choppin, three of the book editors, provide an organizing chapter to open the book. Here, they subject 'equity' and 'discourse' to scrutiny, critiquing the range of ways in which these terms have come to be used, and illuminating their use in this book. Their opening gambit is a discussion of the centrality of discourse, for discourse "is the primary medium of education" (p. 1), followed by a declaration of their shared valuing of equitable pedagogic practice. They trace key issues that have emerged in research related to discourse and equity, with particularly attention to changing discourse patterns in mathematics classroom practice, explicitness in pedagogic discourse, culture dimensions of discourse and equitable discourse practices. These themes, they suggest, are brought together through the chapters and their organization of the book.

Individual chapters in the book focus on particular facets of the relation between equity and discourse, and have been organized by the editors in into three main parts. In the first, chapter authors locate their 'problem' squarely in concerns with equity and inevitably draw in considerations of discourse. Here, concerns for equity in mathematics participation of students from marginalized communities, together with concerns for equitable group-based learning in school classrooms, inevitably require attention to discourse. In the second part of the book, we find the reverse, with chapter authors' problems oriented to discourse that expose issues of equity.

Here the compounding issue of multilingual classroom contexts and the power of a dominant language finds voice. The chapters in each of these first two parts include both theoretical and contextual foci, and provide readers with access to a range of methodological tools in use across studies. The first two parts each have a concluding chapter (by Judith Moschkovich, Part I; by Candia Morgan, Part II) that reflects back on the preceding chapters within it, providing a critique on whether and how the questions of who is learning what and how – that is, the important questions for discourse and equity – are tackled. They further engage with how the messages in the chapters reviewed compound and what questions or issues remain open for further investigation.

The third set of chapters turns attention to implications of the conversations in the first two Parts, including policy implications. There is much in each of these collections for the reader to engage with, reflect on and take forward. Readers will confront theoretical resources from semiotics, sociolinguistics and sociology, as these have been put to use in studies of teaching and learning mathematics, together with issues of identity and power.

Herbel-Eisenmann, Choppin, Wagner and Pimm, together with a group of established researchers in the inter-related fields of discourse and equity contributing chapters, provide a set of scholarly resources that together engage with equity and discourse in mathematics education and their inter-relation. This book is thus a significant addition to the literatures on discourse and mathematical learning on the one hand and on issues of equity on the other, providing insightful and diverse ways in which problems arise, are understood and subsequently engaged. The book provides important reading for researchers in the field who understand, or wish to understand more deeply, that mathematical discourse is constitutive of social relationships and that social relations constitute what comes to be mathematics in and across classroom contexts.

At a time of an ever-increasing pace of migration, with urban classrooms in cities across the world becoming increasingly culturally diverse and multilingual, this book is indeed timely. It offers critical reading for scholars concerned with the role that mathematics education plays in constructing and reproducing both knowledge and social relations.

Jill Adler

Preface

In recent years, the themes of discourse and equity have been widely explored in the mathematics education literature, though not often together. When discourse and equity have been explored together, the goal has typically been to propose ways to increase access to participation in dominant mathematically-based discourse practices, such as argumentation. Although this is a laudable, even necessary, goal, the underlying conceptions of mathematics, equity and discourse around notions of access to dominant mathematical practices ultimately constrain the range of knowledge, ways of being and practices that are deemed legitimate in mathematics classrooms. In this book, we build from expanded views of equity, discourse and mathematics to explore mathematics education in a range of contexts, with the goal ultimately of broadening opportunities for students to become "better persons in their own eyes, not just in the eyes of others" (Gutiérrez, this volume, p. 19).

The work in this book builds from an international conference held in May, 2008 in Rochester, New York, U.S.A., in which over 35 scholars and teachers from six countries spanning both hemispheres and four continents discussed notions of equity and discourse in mathematics education. The quality of the discussion is attributable to the earnest engagement of the participants: Jill Adler, Nancy Ares, Richard Barwell, Mary (Betsy) Brenner, Courtney Cazden, Jeffrey Choppin, Michelle Cirillo, Carol Coles, Liz de Freitas, Helen Doerr, Indigo Esmonde, Ellice Forman, Susan Gerofsky, Rochelle Gutiérrez, Donna Harris, Beth Herbel-Eisenmann, Whitney Johnson, Robyn Jorgensen, Jean Krusi, Lisa Lunney Borden, Lana Lyddon Hatten, Carol Malloy, Candia Morgan, Judit Moschkovich, Mary Catherine O'Conner, Jessica Pierson, David Pimm, Lesley Rex, Tim Rowland, Mary Schleppegrell, Mamokgethi Setati, Angelia Marie Shindelar, Marjorie Spear, Codruta Temple, David Wagner, Gordon Wells and Vicki Zack.

The intellectually and emotionally charged atmosphere at that extraordinary conference set the stage for the work in this book. The conference agenda deviated at key moments, when it was clear that more opportunity was needed to negotiate the meanings of equity and discourse. One discussion in particular provided an opening for a number of scholars to suggest that we expand our focus beyond access to dominant discourse practices and include broader political, social, cultural and historical perspectives. This suggestion was taken up at the time by many of the participants and serves as a central theme in this book. The conference provided the opportunity for a range of perspectives, scholarship and contexts to come forcefully and passionately into contact with one another, in ways that created spaces for many of the participants to question and explore their implicitly held notions of equity, discourse and mathematics. The work in this book thus reflects not only a diversity of perspectives and contexts, but transformations both in those perspectives and contexts themselves and in how they speak to each other.

The conference in Rochester, conceived over a number of years, was funded in 2007 when Jeffrey Choppin, Beth Herbel-Eisenmann and David Wagner received a grant from the U.S. National Science Foundation to support a conference that would bring together scholars who are known internationally for their work in equity or on discourse in mathematics learning contexts. The grant was called *Investigating Equitable Discourse Practices in Mathematics Classrooms*. Conference interaction was structured in the following way. Each of three days started with a plenary panel – 'Diverse perspectives for research on equitable mathematics classroom discourse', 'Working with teachers on mathematics classroom discourse (from inside and outside mathematics education)' and 'Perspectives from stakeholders on opportunities and challenges for reforming mathematics classroom discourse'.

Following each panel, participants discussed and subsequently brought back questions and comments to a plenary discussion. This was followed by 'Research and Action' groups, which were organized around participants' declared interests with the goal of planning research and group action that would address needs in the area of the group's focus. The group foci came from syntheses of interests and issues identified by participants – 'Focus on the student', 'Multimodality', 'Policy', 'Classroom Practice' and 'Approaches to Discourse'. Products of these groups include several conference presentations, a number of co-authored articles, a PME working group and a special issue of the *Canadian Journal of Science, Mathematics and Technology Education*. Less formal collaborations that began at the conference continue to evolve today.

The conference organizers asked David Pimm to join them in co-editing a book that addressed issues raised at the conference and he (thankfully) accepted. Together we identified conference participants whose contributions would complement each other for this purpose. A core group of these chosen contributors gathered in Washington in 2009 to describe their chapter plans, to give feedback to each other, to discuss possible book structures and to write drafts of their chapters. A proposal was submitted in 2010 to Springer for the Mathematics Education Library series, which was accepted.

The primary intended audience for this book includes researchers of mathematics education as well as their graduate students, curriculum decision-makers and teacher educators – particularly those who seek to understand the complex connections between equity and discourse. As most of the chapters are based in empirical research intending to develop theory for interpreting mathematics learning environments, this book is highly appropriate for a research-focused audience. The turn of attention to policy in the last two chapters will be of interest to policymakers, but also to researchers who wish to ground their discussion and theorization in an awareness of the realities facing practitioners. A secondary audience will be researchers interested in issues of discourse and equity more broadly, as our book provides an exploration within a specific and less-commonly attended to school subject.

East Lansing, MI, USA — Beth Herbel-Eisenmann
Rochester, NY, USA — Jeffrey Choppin
Fredericton, NB, Canada — David Wagner
Burnaby, BC, Canada — David Pimm

Contents

Chapter 1
Inherent Connections Between Discourse and Equity in Mathematics Classrooms

David Wagner, Beth Herbel-Eisenmann, and Jeffrey Choppin

Discourse practices warrant the attention of mathematics educators because discourse is the primary medium of education. Though conceptions of mathematics and academic language may often avoid the identification of values, discourse practices can give evidence about whether particular hopes or expectations are being met – no matter if the goal is performance of mathematical procedures, creativity in problem solving or a classroom environment that uses the diversity of voices as a resource. Language, with its associated implicit and explicit actions, is the medium of mathematical development and, consequently, the medium through which equities and inequities are structured and sustained.

We begin this chapter and this book with a statement about what we value – equity in mathematics education. We want it all – equity *and* mathematical learning. We think that equity is something for which most mathematics educators strive. Nevertheless, differences abound in relation to how equity is conceived and how these conceptions are manifested in classroom practices, with consequences for the opportunities to learn for every student. These consequences demand a diligent and sustained focus on equity and discourse.

There is no singular way to conceptualize and explore how discourse and equity are related and how these complex notions play out in mathematics classrooms.

D. Wagner (✉)
Faculty of Education, University of New Brunswick, Fredericton, NB, Canada
e-mail: dwagner@unb.ca

B. Herbel-Eisenmann
Department of Teacher Education, College of Education, Michigan State University, East Lansing, MI, USA
e-mail: bhe@msu.edu

J. Choppin
Department of Teaching and Curriculum, Warner Graduate School of Education and Human Development, University of Rochester, Rochester, NY, USA
e-mail: jchoppin@warner.rochester.edu

B. Herbel-Eisenmann et al. (eds.), *Equity in Discourse for Mathematics Education: Theories, Practices, and Policies*, Mathematics Education Library 55,
DOI 10.1007/978-94-007-2813-4_1, © Springer Science+Business Media B.V. 2012

Consequently, we feel it is important to bring together perspectives from diverse demographic and geographic contexts and widely spread theoretical orientations. Two shared concerns bring the contributors to this book together: an interest in equity and the recognition that equity (and inequity) is expressed, sustained and developed in and through discourse practices. Within these similarities, we note differences in the way that discourse and equity are relatively emphasized. Although every chapter here relates to discourse and equity, some authors have oriented their work around equity and from there were drawn to consider discourse practices, while others have oriented their work around discourse and from there subsequently engaged with equity issues.

In the context of complex notions like 'discourse' and 'equity', both of which can be taken in a variety of ways, definitions of important words, and the ways these words are used, reveal what is valued. The next two sections of this introduction give an overview of our orientations toward the words 'equity' and 'discourse', which were used for determining the scope of the book and to invite contributors. For these central terms we aimed for broad conceptualizations, in order to include diverse perspectives. These two sections are followed by an overview of some essential literature that relates to the connections between discourse and equity, and finally by an overview of the structure of the book.

1 Equity

Attention to equity is part of a larger movement in mathematics education attending to research on sociocultural factors that influence students' experiences (e.g. Brenner 1998; Forman 2003; Gay 2000; Lee et al. 2005). In particular, a considerable body of research on 'funds of knowledge' (Moll et al. 1992) identifies the value of culture and experiences students bring to the classroom to complement the culture and experiences within the classroom and its disciplines. The identification of intersecting (and sometimes competing) systems of knowledge is important to the interrogation of the cultural relevance of mathematics and science instruction. In this regard, Bishop (1988) sought to identify the values embodied in mathematics, while D'Ambrosio (1994) promoted reflection on the effect mathematics and sciences have on society by identifying both horrible and wonderful things that have been enabled by these disciplines. Others have sought to understand and build from the values within given communities by identifying culturally-relevant mathematics teaching (e.g. Brenner 1998; Ladson-Billings 1995b; Warren and Rosebery 1995).

Differences between home and mathematics classroom cultures are also evident in the kinds of practices that are valued, differences which relate to the conceptualization of mathematics. Ethnomathematics research pushes the boundaries of mathematics to include practices that may not have been thought of as mathematical. Barton (2008) made a helpful distinction between "near-universal, conventional mathematics" (p. 10) that is practiced in academic settings internationally (which he called NUC-mathematics) and systems that help people "deal with quantity or measurement, or the relationships between things or ideas, or space, shapes or patterns" (p. 10), which he called QRS-systems. NUC-mathematics or any QRS-system is a form of mathematics set in a particular cultural trajectory. While ethnomathematics is contentious, challenged both by Dowling (1998) and by Vithal and Skovsmose (1997) for instance, ethnomathematicians (e.g. Gerdes 1988) claim that identifying mathematics outside academia serves to counter the inequitable privileging of certain cultures.

Language exemplifies and creates culture and, consequently, the language of instruction privileges culture associated with that language. Thus, attention to language issues at work in multilingual mathematics classrooms (e.g. Adler 2001) is another important part of efforts to make mathematics culturally relevant and responsive. Not only do language choices affect the kinds of discourse privileged in classrooms; they also influence the way mathematics itself can be expressed and thus the way a society learns to address its problems, as shown by Barton (2008).

Cultural relevance is especially important in urban settings, which tend to comprise linguistically and culturally diverse populations. Thus, sociolinguistic research has often taken place in urban settings with high proportions of students from marginalized backgrounds (e.g. Forman et al. 1998; O'Connor et al. 1998; O'Connor and Michaels 1996). However, because inequities relate to the marginalised as much as they do to the powers that sustain the inequities, sociolinguistic research in less diverse settings or in a range of settings is also important. Positioning is always at work in mathematics classrooms, no matter the range or depth of diversity in the context (Wagner and Herbel-Eisenmann 2009).

The positioning of students within classroom interaction is important to their experiences, but students are also positioned in relation both to society and to mathematics. Research on critical pedagogy (e.g. Frankenstein 1998; Gutstein 2006; Powell and Frankenstein 1997) provides both theoretical and empirical accounts of the ways instructional contexts can be designed to address inequities outside the classroom. Positioning students as people who can address issues outside the classroom has repercussions on how they feel in the classroom and how they see themselves in relation to mathematics.

In addition to the attention to equity in the research of mathematics education, equity concerns have also become more prominent recently in professional literature and policy documents. For example, the NCTM standards documents (1989, 1991, 2000), which explicitly or tacitly underpin curriculum and policy in the United States and some other countries, prominently call for greater equity. Equity is the first listed of the six core principles in the U.S. National Council of Teachers of Mathematics's Principles and Standards document (NCTM 2000). However, more recent professional and policy documents in the United States have moved in a different direction: the word 'equity' does not appear in the Common Core State Standards for mathematics (CCSS 2010), nor in the accompanying material published with them on the internet (http://www.corestandards.org/). Additionally, there is a separate three-page document addressing the teaching of English language learners and a two-page document concerned with students with disabilities, rather than equity in mathematics classrooms more broadly. The recent attention to equity at professional and policy levels is sustained and supported by the research, but it also motivates and sustains the research. We have yet to see the interplay between policy, practice and research in relation to the Common Core Standards.

In our view, differences between a student's home culture and the culture of mathematics classrooms are central to structural inequities that exist in mathematics classrooms, particularly since the difference between home and school culture is greater for some than others (Schleppegrell 2004; Zevenbergen 2001b). Home–school differences can be particularly evident when teachers and students are from different cultural, linguistic, socioeconomic or racial backgrounds. Students from marginalised backgrounds, for example, are aware of how racial differences shape the ways they are viewed by teachers and administrators (Howard 2008). More generally, as students, particularly students from marginalized backgrounds, get older, they become more aware of the dissonance between different communities of which they are members, engendering a political awareness that impacts the ways students identify with schools (Lee 2009), but which teachers and schools could use as resources rather than attempt to ignore or suppress. Other research has focused on the deleterious effect of schooling on children from marginalized backgrounds. Research in the U.S., for example, has shown that when background characteristics, especially socioeconomic status, are controlled for, minority children enter school with the same preparation as White students, but lose ground to White students in each year of schooling (O'Connor et al. 2009). Simply focusing on access and achievement will not alter these trends, as the recent experiences in the U.S. with the *No Child Left Behind* experiment have demonstrated. Instead, we side with Gutiérrez (Chap. 2) in pointing to the importance of investigating issues of identity and power and how they play out in classroom interactions, which leads us to elaborate on how we interpret the idea of discourse in this book.

2 Discourse

In relation to discourse, there are at least two sides to our interest in the connections between discourse and equity. First, we consider the ways in which social, mathematical, cultural and political aspects of classroom interactions impact students' opportunities to participate in the kinds of discourse practices that provide access to future resources. Second, we consider the perceptions and practices of educators, particularly the extent to which they view diversity as a resource, as well as that to which they are aware of structural inequities in the ways they perceive and design classroom discourse practices.

We see the word 'discourse' being used to describe how contexts, such as mathematics classrooms, are structured in order to broadly consider how language exchanges embody the diverse social, political, cultural and socioeconomic positions at play. The sociocultural practices related to language use in mathematics classrooms indicate potential mismatches between home-based and school-based practices. Teachers who have a sense of how broad Discourses (Gee 1999) impact the construction of identity and culture can build from students' cultural resources and political awareness to develop mathematical thinking (e.g. Gutstein 2006; Lipka et al. 1998). Our broad approach to discourse opens attention to the structural and systemic influences on educational equity.

There are other shades of meaning related to the word 'discourse'. For example, in much, if not most, of the discussion about discourse in the practitioner literature in North America, 'discourse' refers to oral communication practices in classrooms and tends to ignore the political dimensions of discourse. Because different foci accompany the various ways of using 'discourse', we see value in various approaches to the word in literature aimed at informing research, practice or both. At the very least, in this context of varying meanings for the word, it is important for educators to be clear about how they are using it. We take discourse to comprise a broad range of practices, including reading, writing, speaking and listening, as well as prosodic features of communication and gesture; all are intimately related to the contexts in which such practices are situated and informed by the range of communities in which people participate. Such a sense of discourse, which considers both the practices and the systemic influences on practice, has been detailed by scholars from diverse scholarly traditions, including Michael Halliday, James Gee, Norman Fairclough and Michel Foucault. Although these diverse traditions agree on the importance of understanding discourse, there is tension between orientations, well-documented by MacLure (2003) and within mathematics education by Ryve (2011). These tensions can be generative when scholars engage in conversation around a focused topic, as in this book.

3 Changing Discourse Patterns in Mathematics Classrooms

The resilience of historical patterns of discourse in mathematics classrooms poses concerns with respect to equity; there are strong traditions of practice that privilege certain groups of students, particularly many students from the majority culture in the context of their schooling, in part because these traditions constrain the way students engage with mathematics. Accounts of teacher-directed discourse in which students do little more than provide brief answers to procedural questions and the prevalence of the Initiate–Respond–Feedback (IRF) (Sinclair and Coulthard 1975; Mehan 1979) interaction pattern are well represented in the TIMSS video research (Stigler and Hiebert 1999). These ways of interacting are fairly specific to the context of classrooms. In wider cultural contexts, these kinds of interactions might be considered rude, inappropriate or demeaning. In fact, traditional discourse patterns in mathematics classrooms have been found to be culturally incongruent for some students because of the lack of opportunity for interaction (Brenner 1998), the emphasis on adult authority (Au 1980) and the lack of sensitivity to linguistic concerns (Warren and Rosebery 1995).

Attention to discourse in mathematics education has provided theoretical and empirical descriptions of forms of discourse that seem to provide more equitable and robust conditions for learning. The research is grounded in disciplinary and theoretical traditions that focus on grammatical and lexical features of mathematical discourse (e.g. Halliday 1978; Herbel-Eisenmann and Wagner 2010; Lemke 1990; Mousley and Marks 1991; Pimm 1987; Rowland 2000; Schleppegrell 2004), the sociocultural context in which the discourse takes place (e.g. Lerman 2001; Zevenbergen 2001b, 2005), the sociolinguistic features of discourse (e.g. Bills 2000; O'Connor and Michaels 1993, 1996; Weingrad 1998), and qualities of discourse related to the discipline of mathematics (e.g. Lampert 1990, 2001; Pimm 1987). Research on discourse in mathematics classrooms has been synthesized or summarized elsewhere (Sfard et al. 2001; Lampert and Cobb 2003; Steinbring et al. 1998), but these syntheses have not sufficiently addressed concerns related to equity, diversity and culture. Franke et al. (2007) bring these concerns together, albeit briefly, in relationship to the teaching of mathematics.

Fundamental challenges remain in helping teachers transform their discourse practices to recruit and build better from the diversity of student perspectives and approaches that exist in classrooms. The scant literature that has focused on teachers who attempt to position all students in meaningfully intellectual roles in classroom discourse reports on unusual situations, such as teacher development experiments (e.g. Cobb et al. 1991; Cobb et al. 1992; Yackel and Cobb 1996) or teachers who are considered experts in mathematics education (e.g. Ball 1993; Lampert 1992; McClain and Cobb 1997). Only recently have mathematics education researchers used the tools and concepts of discourse analysis to collaborate with teachers as they teach in their ordinary classrooms (e.g. de Freitas and Zolkower 2009, 2011; Rowland 2000; Staples and Truxaw 2010), including collaborations with teachers that involve action research focused on discourse features (Grant and McGraw 2006; Herbel-Eisenmann and Cirillo 2009; O'Connor et al. 1998; Zack and Graves 2001).

Adler (2001) noted that in any mathematics classroom the teacher has to decide how much to draw explicit attention to language and that this dilemma is exacerbated in multilingual classrooms. Such explicit attention often derives from deficit views of language use (Moschkovich 1999), but need not do so. There is a range of reasons for drawing students' attention to discourse – for example, to support the development of content knowledge, communicative competence or awareness of the discipline's role in society (D'Ambrosio 1994).

Below, we review research on equity and discourse, in order to set the context for the work elaborated in the book. We focus on three themes (elaborated in more detail below) that have emerged as foci for researchers exploring inequitable patterns of participation in mathematics classroom discourse and perspectives that implicate actions for educators attempting to disrupt those inequitable patterns. The three themes are: the difficulties of engaging students in academic forms of discourse; the cultural dimensions of discourse as a means of explaining potential barriers for equitable patterns of participation; the ways teachers can structure interactions to position every student as a contributor to the collective development of mathematical ideas.

3.1 Making Language Practices Explicit

One way to look at discourse with an eye toward equity is to draw attention to the particularities of the mathematics register and the practices of mathematics classroom culture (e.g. Mousley and Marks 1991; Pimm 1987). This research is often intended to help teachers and their students understand and develop competence with the relevant discourses, especially educators supporting students – often those most marginalized in school settings – who have less familiarity with academic and mathematics discourse (Schleppegrell 2004; Zevenbergen 2001b).

With this kind of work, it is important to distinguish between language in mathematics contexts and other contexts. For example, students have probably used the word 'sign' many times outside of their mathematics classroom in reference to things like 'signing' their names on a sheet of paper. In mathematics contexts, however, we talk about the 'sign' of a number, meaning that it is in a positive or negative direction from zero on a number line. We also use the word 'sine' which sounds the same but means something quite different (see Adler 2001). To add further complication, when we write 'sine' or when we use calculators, we write or look for 'sin', which, again, has a completely different meaning outside of the mathematics classroom. (See Herbel-Eisenmann et al. 2009a, for more about this example and other examples of longer stretches of discourse that are different in mathematics classrooms from other domains.) Differences in the way words are used may relate to relatively local ideas, like the meaning of 'sign' and even to ideas central to mathematics, such as the meaning of 'proof' or 'justification'. Explicit discussions about these kinds of nuances in mathematical language may be important so that all students understand the tacit differences.

Further, there are differences between mathematics classroom discourse and other mathematics-related discourse practices. The knowledge, needs and aims of mathematicians are different from but related to the knowledge, needs and aims of mathematics teachers and of mathematics students. These differences appear in the form of their discourse practices, which are specific to their *registers*, their distinct ways of using the natural language in force (e.g. English) to achieve a specific range of functions. Pimm (2007), Barwell (2007) and Herbel-Eisenmann et al. (2010) have argued for the necessity of making this distinction between the mathematics register and mathematics classroom register. It is not appropriate to assume students should be aiming to develop discourse practices that match those of mathematicians. Instead, educators need to consider what form of discourse is the most appropriate for each learning environment.

Coming from linguistic and critical sociological perspectives, some researchers using systemic functional linguistics have argued that it is important for educators to attend to classroom discourse because students need to do more than observe mathematical outcomes. They need to engage in mathematically appropriate (i.e. academic rather than everyday) forms of reading, writing, speaking and listening. It is argued that this kind of explicit engagement in a range of mathematical literacies is especially important when equity issues are considered (Lemke 1990;

Morgan 1998; Schleppegrell 2004). Helping students understand academic forms of discourse provides them access to codes of power (Delpit 1995).

Other research drawing from linguistics and sociology focuses on the social rules and routines present in academic environments. For students to take part in the classroom activities, they must come to follow and understand particular social rules, which are often tacit (Cazden 2001; Voigt 1985, 1989) and not universal in their cultural basis. Classrooms are unlike other social environments because one person, the teacher, is "responsible for controlling all of the talk that occurs while the class is officially in session – controlling not just negatively, as a traffic officer does to avoid collisions, but also positively, to enhance the purposes of education" (Cazden 2001, p. 2). These tacit rules, which are often taken for granted, need to be made explicit to all students, especially students whose home discourses differ most from school discourses. As Cazden and Mehan (1992) explained, students must recognize varying contexts and shift their language use multiple times throughout the day, and even within a particular lesson. For example, the ways in which students are expected to participate in small groups varies considerably from the ways they are expected to participate during a teacher's lecture; these expectations come from each other as well as from the teacher. Different interaction patterns involve various methods used to control classroom discourse (see, for example, Edwards and Mercer 1987) and can be more or less aligned with the discourse patterns of other communities of which students are members.

Because many of these routines and rules are tacit, the identities of the participants play a significant, yet often overlooked, role in classroom discourse practices (Cazden 2001; Evans 2000; Hannula 2002; Heath 1983). A focus on both classroom discourse and identity is connected to equity issues for at least two reasons. First, there is an increased emphasis on teaching mathematics to every student[1] (NCTM 1989, 2000) and a concern about achievement gaps between sociocultural groups (Lubienski 2002; Lubienski et al. 2004). Second, there are often major differences between the demographics of the teaching force and those of the student population, with the teaching force disproportionately constituted from dominant groups while the students are increasingly diverse. In the U.S., for example, over four-fifths of elementary teachers are White, while non-White students will soon be in the majority (USDOE 1998).

Discourse studies on literacy practices have highlighted the importance of perceived status (by both teachers and peers) and one's identity as they influence both the dynamic rituals and routines of small- and whole-group interaction (Lewis 2001). Similarly, it is likely that informal social and home discourses could impact the ways that students are taught to view and engage in argumentation (O'Connor 1998), something that is viewed as an important mathematical process in policy documents (e.g. NCTM 2000).

[1] In the NCTM Standards documents, the earlier version used the words "all students". In more recent publications, however, the words "every student" were used instead. We see this as a potentially important shift, because it indexes a move from a perspective of equality (in which all students get access) to a perspective of equity (in which close attention to each student can help educators to provide what each student needs).

In order to address disparities, Morgan (1998) concluded her critical discourse analysis of student mathematical writing by saying that students need to be drawn into critical discussion about their discourse, offering as a model Fairclough's (1992) critical language awareness, which is typically done in language-oriented classes. When Wagner (2007, 2008) explored ways of doing this, students were fascinated by and engaged with certain conversations about discourse, while apparently unmoved by others. The mixed results of Wagner's work in one context warrants further exploration, especially because students' concerns are usually not explored by researchers. Special issues on discourse and equity of the *Canadian Journal for Science, Mathematics and Technology Education* (**11**(3), edited by Esmonde and Moschkovich) and the *Journal for Research in Mathematics Education* (**41**(0), edited by a panel chaired by Gutiérrez) partially address this need.

3.2 Cultural Dimensions of Discourse

The discipline of mathematics and its development in classrooms sits in a wider context that privileges certain values and world-views above others: thus, culture is closely related to discourse. Mathematics content and instruction "should enable children to build from their existing cultural base in mathematics" (Brenner 1998, p. 215). School, the first large institution in which students are expected to participate individually and publicly, involves multicultural encounters with both teachers and students belonging to diverse groups differentiated by variables such as age, social class, gender, race and ethnicity (McGee Banks and Banks 1995). Thus, teachers must respect and seek forms of student participation that are consonant with children's everyday ways of thinking and living (Moll et al. 1992; Moschkovich and Brenner 2002). This point has significant implications for mathematics instruction. The traditional structure of mathematics classes constrains opportunities for diverse cultural traditions to serve as classroom resources. Brenner (1998) states that, "there is substantial evidence that the participant structure of a traditional classroom, that is, the roles and responsibilities assigned to the different persons, can act as an inhibiting factor to children who come from a culture that stresses different participant structures than those found at school" (p. 215).

As noted above, traditional mathematics instruction can be culturally incongruent because of the limited ways in which patterns of interaction tend to draw on cultural and linguistic resources. We believe that instruction needs to provide greater opportunities for interactions between participants, interactions in which students' cultural resources can be used to communicate particular perspectives or solutions that contribute to the collective negotiation of knowledge. Teachers should attempt to gain an in-depth understanding of their students' backgrounds and the relationship between their cultures and their learning (Malloy and Malloy 1998). For example, getting to know students' backgrounds can allow curriculum material modification in order to make tasks compelling and applicable to students' experiences (see El Barrio 2009); it can also help teachers assist students to use mathematics in critical ways (Gutstein 2006).

Increased interaction allows household 'funds of knowledge' to serve as a resource for students (Moll et al. 1992). Such 'funds' should not merely reproduce home-based cultural practices related to specific professions, but rather household and other sources of knowledge should be drawn upon so that "student experience is legitimated as valid and classroom practice can build on the familiar knowledge bases that students can manipulate to enhance learning in mathematics [and other content areas]" (González 1995, p. 240). As Gay (2000) pointed out:

> teachers should not merely make girls talk more like boys, or boys talk more like girls, or all individuals within and across ethnic groups talk like each other [...] Instead [teachers] must be mindful that communication styles are multidimensional and multimodal, shaped by many different influences. Although culture is paramount among these, other critical influences include ethnic affiliation, gender, social class, personality, individuality, and experiential context. (p. 109)

Gay's 'culturally responsive teaching' articulates ideas upon which educators can draw to consider how to connect deeply with students.

The work of Au and colleagues in Hawaii (Au 1980; Au and Jordan 1981) and Warren and Rosebery (1995) among Creole children of Haitian descent demonstrates how instruction that explicitly recognizes and builds from students' cultural resources can promote the development of academic knowledge. These students' unique position straddling a border between cultures, a position that has also be characterized as a third space (Gutiérrez et al. 1995), affords them and their peers a view of the cultural aspects of mathematics discourse. Ladson-Billings' (1994) study of teachers recognized as effective by both community members and administrators found that these teachers made great efforts to build aspects of students' community life into their classrooms as a means to help students learn. Similar work shows this connection in other contexts, especially for Aboriginal students (e.g. Aikenhead 2002; Orr et al. 2002; Tompkins 2002).

3.3 *Structuring Equitable Discourse*

The differences between home cultures and school may paint a bleak picture for mathematics education. However, some scholars have investigated discourse to propose patterns of interaction that depart from the traditional I–R–F structure. For example, O'Connor and Michaels (1993, 1996) documented specific linguistic practices that serve to socialize students into mathematical aspects of argumentation. They detailed the impact of these moves in terms of the creation of participant frameworks, which provide opportunities for students to take part in particular types of complex thinking by means of "taking on various roles and stances within recurring social contexts that support [...] intellectual give-and-take and its proto-forms" (1996, p. 64). Such *revoicing* functions to align students' explanations with academic content and with each other, attributes ideas to students and ultimately serves to portray students as competent mathematical thinkers. Moschkovich (1999) described how a teacher's use of revoicing incorporated the contributions of bilingual students into mathematical discussions. The teacher moved beyond a focus on language development to engage students in aspects of discourse recognized as central to the learning of mathematics, providing them with opportunities to develop competence in academically valued practices. The teacher thus played a prominent role in "uncovering the mathematical content in student contributions and bringing different ways of talking and points of view into contact" (p. 11), as a means of helping students experience academic excellence.

Just as significant as the investigation of teacher moves such as revoicing are accounts of discourse practices that limit student agency in general or the agency of particular student groups. For example, in Morgan's (1998) discourse analysis of student mathematical writing, she noted grammatical functions inherent in the discourse that obscure the agency of participants. This obfuscation mirrors the hidden agency in mathematics textbooks (Herbel-Eisenmann 2007) and the way oral mathematics classroom discourse positions students into complicit roles (Herbel-Eisenmann and Wagner 2010; Wagner and Herbel-Eisenmann 2008). Thus, there is a need for explorations of efforts to change discourse, which, we suggest, might follow the methodology for critical mathematics education research described by Skovsmose and Borba (2004) or by others in the collection edited by Valero and Zevenbergen (2004). Within such reflection-intensive work, we would also support the use of theoretical tools from discourse-related disciplines, including linguistics and cultural studies.

4 Bringing These Perspectives Together in This Book

To bring together perspectives oriented to discourse and equity in the context of the research described above, we have structured this book to include in Part I examples of work that starts from an orientation to equity and, in Part II, examples starting with a discourse orientation. The work that starts with equity indicates that a focus on equity draws attention to discourse. We think this shift is inevitable in any discipline or situation (mathematics education or not), because of the way human relationships, whether equitable or not, express themselves in the discourse and because change in relationships operates through discourse.

The work that starts with discourse shows that in mathematics education discourse structures have implications for equity. In this case, we only claim this connection for the context of mathematics education, though we recognize that the phenomenon may occur in other situations (see, for example, Heath's 1983 classic study and intervention in elementary literacy). Particularities of the language practice in mathematics and in mathematics classrooms are oriented to generality and abstraction. Thus, students, though connected to the mathematics in their learning, may seem insignificant or ignored. We note that ignoring differences between people can be either freeing or oppressive, but nevertheless with either extreme there are implications for equity.

The first two sets of chapters, then, move from equity to discourse and from discourse to equity, respectively. Within each set, the first chapter is strongly theoretical, the second focuses on a particular context but with strong orientation to theory, the third and fourth focus on the context of the research, and the fifth looks across the set to reflect on the relevant move (from equity to discourse or from discourse to equity). These two sets of chapters are followed by two chapters that consider policy implications, and a response to the entire conversation comprising the book.

The first group of chapters (Part I) begins with Rochelle Gutiérrez noting differences in the way educators conceptualize equity, which she shows to encompass access, achievement, identity and power issues. Following this, there are three chapters (Chaps. 3, 4, 5) describing work that was oriented to equity but that includes discourse implications. In the first, Robyn Jorgensen employs a Bourdieuian analysis to explore the 'synergy' between the culture of school mathematics and the cultural practices students bring to school, especially Aboriginal students in an Australian context. In the second of the group, Indigo Esmonde develops language for describing equitable classroom group structures for mathematical exploration. In the third, David Wagner and Lisa Lunney Borden reflect on their ethnomathematical research amongst Mi'kmaw First Nation communities on Canada's east coast, noting positioning issues in the discourse as they strove to structure respectful relationships. The section ends with a reflection chapter written by Judit Moschkovich, who looks back across this section to show how concern for equity necessitates attention to discourse.

The second set of chapters (Part II) begins with linguist Mary Schleppegrell's articulation of how linguistic tools might be used to consider particular language choices in mathematics classrooms, thereby examining the construction of mathematics offered to students, as well as the positioning of students in relationship to mathematics. She identifies questions and issues related to equity for which examination of language choice might help researchers. Following this chapter, there are three chapters (Chaps. 8, 9, 10) describing work that was oriented to discourse and that exposed issues of equity. In the first, Mamokgethi Setati reflects on the research she has done over the past decade and shows how the dilemmas teachers face in multilingual classrooms relate to equity beyond the classroom. In the second, Richard Barwell illustrates the idea of 'discursive demands' as a way of thinking about some aspects of the double challenge faced by English as a second language learners in the U.K. attempting to learn mathematics. In the third, Beth Herbel-Eisenmann discusses a dilemma of telling, one which relates to one's right to call on one's authority and to control social (and mathematical) aspects of others' work, that emerged as she led teachers to attend to their discourse. A final reflection chapter by Candia Morgan looks back across this section to highlight how mathematics learning discourse in particular comprises inherent equity questions.

The third set of chapters (Part III) turns attention to implications of the conversations in the first two sections, in particular policy implications. First, Donna Harris and Celia Anderson consider the ways that policy shapes opportunities for teachers and students to engage in more demanding – and valued – forms of discourse. Second, we draw on conversations with selected mathematics education stakeholders to consider further implications of research that relates to equity and discourse.

A key specificity of this book is that of the work under discussion took place in countries where English is the primary, if not the only, national language of teaching and learning. Following his observation that his book is written in English, Barton (2008), in *The Language of Mathematics*, goes as far as to claim, "To the extent that mathematical ideas differ between languages, the reflexive principle means that the ideas in this book would be different if they were written in another language" (p. 11). As David Pimm points out in his Afterword, we are far from understanding the effects of the current substantial hegemony of English in many educational situations. In consequence of the foregoing, there is more than a significant geographic constraint in terms of the location of the research sites indicated by the foregoing observations. Nevertheless, despite these important limitations, this book assembles a diversity of perspectives in order to address the intertwining of discourse and equity issues in the context of mathematics education.

Part I
Equity Concerns Draw Attention to Discourse

Chapter 2
Context Matters: How Should We Conceptualize Equity in Mathematics Education?

Rochelle Gutiérrez

Contexts have always mattered to me. Perhaps it is because I was raised to believe that communities shape and support individuals into the beings they become. Some contexts bring out the best in me, while others hide my strengths. Considering my worldview, it makes sense that my research would pay particular attention to contexts.

In my research, I do not strive for the empirical findings to be generalizable to all students or even all U.S. students. My focus has always been to document successful learning environments for students who have been marginalized by society, highlighting the origins of such learning environments – be they personal or institutional. I do so for two main reasons: (1) as an existence proof to those in doubt that these environments and their associated student outcomes can be created; (2) as a means for informing how we might build more such contexts for learning. By marginalization, I mean through processes such as racialization, classism, sexism and language bias. However, that is not to say that many of the foundational pieces of these successful environments are not applicable in settings where the students are white and/or middle/high income.

Contexts matter for a number of reasons. A focus on context helps remind us that no category of teachers or students (urban students, African American students, Latina/o students, even female, bilingual Latinas born in the U.S.) is homogeneous. In fact, our beliefs, our lived experiences, our knowledge bases and our agendas all influence how we 'perform' in a given setting. All good teachers focus on context. They recognize the fact that among other things, a student's mathematical thinking is grounded in the kind of problem presented, how that student is positioned in the classroom with respect to others (De Avila 1988; Forman and Ansell 2002), the norms of interaction (Seeger et al. 1998) and the tools available to express one's ideas (Khisty and Viego 1999; Moschkovich 2007a).

R. Gutiérrez (✉)
Department of Curriculum and Instruction and Latina/Latino Studies,
University of Illinois at Urbana-Champaign, Champaign, IL, USA
e-mail: rg1@illinois.edu

B. Herbel-Eisenmann et al. (eds.), *Equity in Discourse for Mathematics Education: Theories, Practices, and Policies*, Mathematics Education Library 55,
DOI 10.1007/978-94-007-2813-4_2,

For me, a focus on the context of learning also serves as a humanizing tool in mathematics education research. It moves us away from a kind of objectified way of knowing something (e.g. students or the 'one' path to equity). And, contrary to what the larger public, many alternative certification programs and some mathematicians think, mathematics teaching is too complex to be reduced to a list of basic skills or even strategies that can be followed by any college graduate. So, while it is important for mathematics educators to present their research in ways that are accessible to policymakers (Lubienski 2008), giving voice to the contextual factors that enable or constrain learning in a given situation is equally important. Richer descriptions of educational settings and their origins also are more likely to move away from a U.S.-centric perspective and towards a more global reality in reporting mathematics education research.

My work is deeply grounded in sociocultural theory, drawing on the notion that learning is intricately connected to the contexts in which it occurs (Lave 1991; Lave and Wenger 1991; Cobb 2000; Atweh et al. 2001). We see this most clearly in research that has considered out-of-school versus in-school mathematics performance (Nunes et al. 1993; Civil 2006). Almost at the flip of a switch, highly competent street vendors are unable to complete similar mathematical problems when imported into a 'school math' context. Like Franke and Kazemi (2001), who seek to "capture the evolutionary character of teacher learning rather than the more static characteristics" (p. 56), I aim to document the nature of effective teaching and learning contexts, not just their distilled 'characteristics'.

Most members of the mathematics education research community would agree that equity is a valued goal, maybe even the reason behind their research. However, much less consensus arises when the question is raised: *how do you think we should address equity?* Increase teacher content knowledge, create more multicultural curricula, develop professional learning communities, exert greater control over school policies or partner universities with local schools are just a few of the strategies that might start the list. For the most part, highlighting (successful) contexts is not likely to be an answer. Yet attending to context is key for equity purposes. In this chapter, I will unpack a few contexts in which I have conducted research and what they have revealed to me about equity along four dimensions. Then, I conclude with ways in which teaching and learning contexts, especially successful ones, might play a larger role in future research. The contexts I will explore include: nine U.S. high schools, one successful teacher community, 23 teacher candidates and the achievement gap.

1 Framing Equity

I begin with a definition of equity, partly because it is critical to how we might explore successful contexts and because so many definitions of equity exist. Equity means fairness, not sameness. So, when we look for evidence that we are achieving equity, we should not expect to find that everyone ends up in the same place. Nearly a decade ago, I argued (Gutiérrez 2002a) that at a basic level, equity means "the inability to predict mathematics achievement and participation based solely on student characteristics such as race, class, ethnicity, sex, beliefs, and proficiency in the dominant language" (p. 153). I argued for a focus on the dominant interpretation of this meaning as well as a critical one (something I will discuss later), and how equity could relate to the sustainability of our planet. It was important for me at the time to consider not just learning outcomes as they relate to a schooling context, but also to learning outcomes that relate to life and our relationships around the globe. I would like to elaborate on that definition to include four dimensions: access, achievement, identity and power.

Access relates to the tangible resources that students have available to them to participate in mathematics. These resources include such things as: high-quality mathematics teachers, adequate technology and supplies in the classroom, a rigorous curriculum, a classroom environment that invites participation, reasonable class sizes and supports for learning outside of class hours. The Access dimension reflects the predominant equity mindset of math educators in the 1980s that students are affected by their 'opportunity to learn' and continues today in more nuanced forms (Nasir and Cobb 2007). However, a focus on access is a necessary but insufficient approach to equity, in part because equal access assumes sameness.

Beyond opportunities to learn, we also care about student outcomes, or what I categorize as *Achievement*. This dimension is measured by tangible results for students at all levels of mathematics. Achievement involves, among other things, participation in a given class, course-taking patterns, standardized test scores and participation in the math 'pipeline' (e.g. majoring in mathematics in college, having a math-based career). Moving from mere access to achievement is important when considering that there are serious economic and social consequences for not having enough math credits to graduate from high school, not scoring high enough on a standardized achievement test to gain acceptance to college or not being able to major in a math-based field that can confer a higher salary and greater prestige in society. The achievement dimension was most prominent in the late 1980s and early 1990s, when a greater emphasis was placed on standardized test scores and continues today into the more narrowly defined 'achievement gap', something I will discuss later in this chapter.

However, because there is a danger of students having to downplay some of their personal, cultural or linguistic capacities in order to participate in the classroom or the math pipeline, and because some groups of students historically have experienced greater discrimination in schools, issues of *Identity* have started to play a larger role in equity research in mathematics education (de Abreu and Cline 2007; Martin 2000, 2007). In my view, students should be able to become better persons in their own eyes, not just in the eyes of others. For most mathematics educators, identity issues might include understanding mathematics as a cultural practice in ways that might further develop the appreciation of one's 'roots'. Examples of this approach are present in the ethnomathematics program (D'Ambrosio 2006). But we cannot stop there, as identity is much more than just one's past.

More centrally, the identity dimension concerns itself with a balance between self and others. A window/mirror metaphor is useful here: that is, students need to have opportunities to see themselves in the curriculum (mirror), as well as have a view onto a broader world (window). For example, using mathematics to analyze social justice issues might offer a mirror to students who have been marginalized by society, while it provides a window to students who benefit from the status quo. Identity incorporates the question of whether students find mathematics not just 'real world' as defined by textbooks or teachers, but also meaningful to their lives. It includes whether students have opportunities to draw upon their cultural and linguistic resources (e.g. other languages and dialects, algorithms from other countries, different frames of reference) when doing mathematics. As such, we need to pay attention to the contexts of schooling and to whose perspectives and practices are 'socially valorized' (de Abreu and Cline 2007; de Abreu 1999; Civil 2006). The goal is not to replace traditional mathematics with a pre-defined 'culturally relevant mathematics', but rather to strike a balance between the number of windows and mirrors provided to any given student in his/her mathematics career.

However, even if students have access to high-quality mathematics, achieve a high standard of academic outcomes as defined by the status quo and have opportunities to 'be themselves and better themselves' while doing mathematics, it is not enough to call it equity if mathematics as a field and/or our relationships on this planet do not change. As such, a final piece of equity involves *Power*. The Power dimension takes up issues of social transformation at many levels. This dimension could be measured (see Gutiérrez 2002a for a more developed argument) in terms of:

- voice in the classroom (e.g. who gets to talk, who decides the curriculum) (Morales 2007; Zevenbergen 2000; Adler 1998a);
- opportunities for students to use math as an analytical tool to critique society (e.g. exploring 'risk' in society) (Mukhopadhyay and Greer 2001; Skovsmose and Valero 2001; Gutstein 2006);
- alternative notions of knowledge (D'Ambrosio 2006);
- rethinking the field of mathematics as a more humanistic enterprise (e.g. recognizing that math needs people, not just people need math).

For the most part, Access and Achievement can be thought of as comprising the *dominant axis*. By 'dominant', I mean:

> mathematics that reflects the status quo in society, that gets valued in high stakes testing and credentialing, that privileges a static formalism and that is involved in making sense of a world that favors the views and perspectives of a relatively elite group. (Gutiérrez 2007a, p. 39)

These are the components students will need to be able to show mastery of in the discipline as it is currently defined and to participate economically in society. This axis, where access is a precursor to achievement, measures how well students can play the game called mathematics.

On the other hand, Identity and Power make up the *critical axis*. By 'critical', I mean:

> mathematics that squarely acknowledges the position of students as members of a society rife with issues of power and domination. Critical mathematics takes students' cultural identities and builds mathematics around them in ways that address social and political

issues in society, especially highlighting the perspectives of marginalized groups. This is a mathematics that challenges static formalism, as embedded in a tradition that favors the West. (p. 40)

The critical axis ensures that students' frames of reference and resources are acknowledged in ways that help build critical citizens (Skovsmose and Valero 2001). In some sense, identity can be seen as a precursor to power. This axis builds upon the idea that mathematics is a human practice that reflects the agendas, priorities and framings that participants bring to it. As such, a diverse body of people is needed to practice mathematics, not just to build a twenty-first-century work-force, but also so that they might participate democratically. Moreover, mathematics needs a diverse body of people so that the field can sustain itself in the most vibrant way possible.

To be clear, all four dimensions are necessary if we are to have true equity. Learning dominant mathematics may be necessary for students to be able to analyze the world critically, while being able to analyze the world critically may provide entrance into dominant mathematics. It is not enough to learn how to play the game; students must also be able to change it. But changing the game requires being able to play it well enough to be taken seriously. As researchers concerned with equity, we must keep in mind all four dimensions, even if that means that at times one or two dimensions temporarily shift to the background. A natural tension exists between mastering the dominant frame while learning to vary or challenge that frame. As such, access, achievement, identity and power are not going to be equally or fully present in any given situation. For example, teachers cannot be expected to address power issues every day in the classroom in ways that are meaningful to every student. Similarly, when identity or power issues are being brought to the surface, at times the connection to mastering dominant mathematics may take a lower priority. The goal is to attend to and measure all four dimensions over time.

2 Equity in Teaching and Learning Contexts

Given this broader definition of equity, we might ask ourselves: *how do access, achievement, identity and power play out in different contexts? Which contexts matter? How do they matter for promoting equity?* In this section, I will unpack a few contexts in which I have conducted research and argue what they have revealed to me about equity. In each of these contexts, I ask: what is the nature of this context and how does it contribute to our understanding of equity?

2.1 Nine U.S. High Schools

I have always believed we learn best from understanding 'success' cases. In that vein, I first began my research trajectory with this question: *What is the nature of a public school that propels its students not only to take more mathematics than is required by the district, but also to show significant gains in standardized achievement?* Steeped in 'opportunity to learn' theories, my first cut was to take an institutional/policy analysis, focusing on tracking as it affected students' access (Gutiérrez 1996). I drew upon the Longitudinal Study of American Youth (Miller et al. 1992), a data set following students from grades 7–12. Using hierarchical linear modeling to capture the effects of students nested within schools, I sorted the 52 schools based upon overall student gains in mathematics, course-taking patterns and differentiation within student outcomes. From the larger data set, I chose nine U.S. high schools that were non-selective and serving a large proportion of Latina/o, African American and/or working-class students. Four of these schools were chosen for their clear student gains and signs of success; four other schools were chosen for negligible signs of success with little or no gain (e.g. less than 50% of students at the school reached the second year of Algebra by grade 12); one school was chosen to represent middle-of-the-road schools. My goal was to understand the nature of these schools and their accompanying success (or lack thereof). I supplemented the quantitative student data with teacher questionnaires, teacher interviews and school documents.

Though much research at that time had focused almost exclusively on the practices and outcomes of individual teachers or school-wide cultures, I changed the contextual frame to consider teacher community in relation to institutional issues. For me, a single teacher was not the appropriate context for getting at broader notions of equity. Moreover, a school-level analysis was likely to minimize the role of subject matter in teachers' everyday work commitments (Gutiérrez 1998; Stodolsky and Grossman 1995; Talbert 1995; Siskin and Little 1995). I was interested in the four schools where a large proportion of their students were excelling in mathematics and where that distribution was spread out over the entire student body. For me, that had to involve more than one maverick teacher or a silver-bullet policy. The mathematics department seemed a useful unit of analysis.

What distinguished the effective math departments from the ineffective ones? Tracking was not the pivotal policy. In fact, two of the four successful schools had tracking policies in place, with support structures to push adolescents towards higher-level courses, and half of the ineffective schools were de-tracked. The number of formal departmental meetings, years and degrees of staff members, math/science magnet designation and overall school culture also were not key to distinguishing success. Instead, the effective departments stood out as different from the ineffective ones in four main aspects of their organization and culture. They had a rigorous and common curriculum, commitment to a collective enterprise, commitment to students and innovative instructional practices.

A rigorous and common curriculum meant there were very few lower-level mathematics courses in which students could get lost or bored. In fact, students were offered little choice in the kinds of courses they could take, as streamlined paths led to the most advanced courses and 3-year minimum requirements for graduation were implemented. Additional courses were created to get students back on track or help them double-up courses in a given year, so they did not lose sight of the end goal. In their curricular design and their course requirements, these effective math departments presented to students a culture that taking higher-level math courses was not only expected, but also just the norm.

The second component to these effective math departments was a commitment to a collective enterprise. That is, unlike the norms of privacy found in many schools, teachers in these departments regarded themselves as part of a community of practice (Lave and Wenger 1991; Wenger 1998), learning from and with colleagues. One of the first signs of this collective priority was the practice of rotating teachers' course assignments, so that no single teacher owned all of a single category of students (e.g. freshmen, seniors, honors students) or subject matter (e.g. all geometry classes, all algebra classes). Teachers explained that rotating the courses meant that they not only had a chance to get a broader sense of the mathematics curriculum (e.g. reminding themselves of how algebra is the foundation of calculus), but it also allowed for repeat students – ones who were in a given teacher's class for more than 1 year. The impact of these repeat students was that teachers often had to think twice about judging a student as either innately competent or incompetent, as they noticed that some students were just late bloomers, going through family issues or better at certain topics than others.

This course rotation also led to more teachers discussing their work and sharing lesson plans. While many of the ineffective departments could be described as operating under an 'independent contractor' mode, the effective departments relied upon each other for professional development. At times, they attended workshops and courses together based upon the subject matter taught, while at others, they required individual teachers to report back to the group on events they had attended. These departments could be described as having collective autonomy in the sense that they did not conduct all business as a whole group. Rather, they had a common vision of what they were trying to accomplish and used frequent discussions and activities to address their goals.

The third component was a commitment to all students. More than just a slogan, this commitment came through in teachers' actions. For example, rather than the deficit frames or stereotypes held by members of the ineffective departments, teachers held constructive conceptions of students (e.g. as creative, smart) and held them accountable to high expectations. Partly related to the 'repeat' students that teachers mentioned, they held flexible conceptions of the learning process (e.g. that not all learning could be easily measured or that maturity contributed to proficiency). They also shared the responsibility for learning, seeing it as partly their role to motivate students to want to learn.

The fourth component distinguishing the effective math departments from the ineffective ones was innovative instructional practices. In terms of instruction, while I found ineffective and/or traditional teachers in effective departments and successful teachers in ineffective departments, they were exceptions rather than the rule. Overall, while teachers in the effective math departments for the most part continued to lecture, they moved beyond worksheets and practice of basic skills. Moreover, as a group they attempted to make the mathematics relevant to students' lives, partly by offering choices of topics for larger projects. Some such projects included basketball standings, ages of actors/actresses at the time of receiving an Oscar and

African American voter registration. Technology was also more prominent in these effective departments than in the ineffective ones. The majority of teachers used graphing calculators to model concepts and to help students see dynamic patterns or 'the bigger picture'. What is more, students were expected to work in groups – partly to attend to the personal need for students to be engaged with peers, but also to encourage reasoning and conjecturing.

Although I have outlined the four components individually here, no single component would be enough to create the success these departments saw. More likely, the effects were synergistic – building off of each other. I termed this departmental culture 'Organized for Advancement' (Gutiérrez 1996), suggesting it involved a conscious 'stance' (Cochran-Smith and Lytle 1999) on the part of teachers to organize themselves and structure their work in ways that advocated for students and their learning above everything else. That is, it is not the mere presence of these components as resources for teachers that matter; it is also the meanings that emerge for teachers and students as these resources are put into use in local contexts (Adler 2001).

From an equity standpoint, three of the four dimensions are highlighted: access, achievement and identity. More specifically, when mathematics departments organize their formal and informal policies, courses, interactions and supports for students in ways that promote high standards, students not only gain access to high-quality mathematics, they tend to achieve in ways that relate both to broader participation and to test scores. When students are offered the opportunity to choose their own topics for projects, to a certain extent they are invited to express their identities and/or draw upon their cultural resources. What was clear to me at the end of this study was that although I could distil the results of the nine schools into a set of four characteristics that distinguished effective from ineffective mathematics departments, I was only scratching the surface. I needed to explore in greater depth the nature of a single math department, partly to understand better the dynamics involved. Lastly, while I was convinced that these OFA mathematics departments were addressing access and achievement, I was skeptical that identity and power issues were sufficiently acknowledged (Gutiérrez 1999).

2.2 A Successful Teacher Community

The focus shifted in my next study to ask not only what was the nature of a successful mathematics department, but also how was this teacher community created and sustained? Again, I continued to search for answers to the question: *what does this community reveal about equity?* This math department was situated within a school that served 67% Latina/o students, 15% African American and with 98% qualifying for free lunch. Their success was measured by: students taking more than the required number of math courses while in high school, a large number of students in calculus (30 in 1996; 42 in 1997; 61 in 1998; 80 in 1999), calculus classes reflecting the broader student body (e.g. with respect to race/ethnicity, class, language and school success) and 80% of the calculus students college-bound (Gutiérrez 2002b). The following vignette attempts to capture the school context.

> We enter Union High School through the backside of the building and pass through a set of metal detectors and two armed Chicago Police officers standing post. Students (primarily working class and Latina/Latino) no longer enter through the front because it faces a main road that provided access for a shooting in the 1980s. Streams of students with large red identification tags swinging from their necks push past each other to get to their classrooms and to socialize with their friends. Students are ushered through the halls by security staff in red shirts and teachers (mainly white and middle class) who also display identification tags. A look at school test scores indicates many of the freshmen are several years below grade level in skills, especially mathematics and English. Union is what the media often portrays as the degradation of public schools in the inner city.
>
> We might expect this school to offer an array of low-level ('business math', 'consumer math') courses, a watered-down curriculum with perhaps one AP calculus where those few students who make it through the public school system are still interested in college and a possible career in math. Instead, we find three full calculus classes.
>
> Each teacher has his own personal style. One has a dry sense of humor, cracking jokes with his students and then quickly getting down to business. Another has a soothing voice accompanied by energetic presentations and passion about mathematics. Still another has a relaxed and youthful air to being with students who are close to him in age. In all three classrooms, we see Latina/o students (primarily) with some African Americans and just a few whites all working in groups, communicating and arguing about mathematical concepts and strategies for approaching problems. They alternate between Spanish and English language, between graphing calculators and pencil/paper forms, between time spent at their desks and at the chalkboard or their small-group white boards, between their textbook written by Harvard professors and worksheets made by their own teachers, between understanding mathematics as the "forest" (big picture/concepts) and the "trees" (details/symbols) and learning from examples that incorporate students' and teachers' lives – all with the goal of understanding the meaning of derivative and integral.
>
> In each class, teachers are walking around to groups of students posing provocative questions and/or providing feedback for student work. Mostly, the teachers project a facilitator role, encouraging the students to help each other. Students pick up on this fact and are getting up from their tables to confer with other groups before returning to share the information obtained or to tutor other students when everyone in their group has reached an answer. These classes could not be described as quiet. Rather, they have the "hum" of intellectual activity that would make most teachers proud. And, with forty percent of the school's senior class present in these three calculus classes, who wouldn't be? These classes reflect both some of the goals that NCTM has put forth in the Standards and the formats used in countries where math achievement is high.

Through classroom observations, teacher and student interviews and an analysis of school documents, the strong role of teacher community came through. In the words of one teacher:

> I think actually individual really good teachers help some kids that wouldn't make it otherwise, but I think the task of a department or of a school is to build up a community, a spirit, a plan that makes it broader than just one individual teacher, you know. And I think that may be the key lesson of what we've done at Union, that it's bigger than one teacher. And the power of a bunch of teachers working together is like greater than, the whole is greater than the sum of its parts.

In fact, only through community were teachers able to support students cognitively and emotionally in ways that advanced them to calculus.

Like the math departments that were Organized for Advancement, this department rotated its course assignments so that no single teacher owned a set of students or topics. The lack of teacher tracking in this successful teacher community was less a result of a school policy and more reflective of the stance that teachers took to create more democracy and opportunities for learning among themselves. Teachers could also be found sharing and discussing curricular materials; communicating and reflecting on students and their teaching; reinforcing to each other that all students can learn calculus; relying upon each other for professional development and support for students. Like the OFA departments I had studied, a key feature of this successful context lay in teachers placing student needs, not just mathematics, at the center of their work.

While the broader mathematics education community has embraced the idea of 'Lesson Study' (Fernandez and Yoshida 2004; Crockett 2002) and 'Communities of Practice' (Stein et al. 1998; Franke and Kazemi 2001; Sherin and Han 2004), it is important for equity purposes to consider whether teacher community should be an end in itself (as a universal model of professionalism and growth) or a means to something larger. In fact, the vision of student empowerment, not just professionalism, drove the norms and practices of this teacher community. In the words of the department chairperson:

> More than anything we provide a vision for kids [...] having them believe in themselves as a group, having them be able to do math as a group, having them believe they can go to college as a group and then at a whole 'nother level, um, it's like a political level [...] Organizing, I mean, I, I mean, at some level my way of teaching tries to organize them to be actors rather than acted upon.

As such, we learn that for equity purposes, the guiding mission of a community of practice may be as important, if not more so, than its presence.

Upon further exploration, this successful teacher community could not easily be distilled into a set of static characteristics without regard to how the community developed or was sustained through threat. A look into the history of this community of practice showed it was built partly on the biographies of the most veteran teachers (many of whom held identities that were marginalized in society), partly on a university partnership that provided professional development and partly on strategic recruitment and socialization of new members over a period of 10 years (Gutiérrez and Morales 2002). When teachers' practices and beliefs were threatened by a new principal who sought to focus staff on basic skills, the community's strong commitment to students and a reform curriculum, coordination of courses, mentoring of new teachers and joint lesson planning allowed them to continue many of their practices without administrative support or sanction. Their community of practice had effectively helped them subvert the system so they could continue to be advocates for students. As such, this study highlighted the importance not only of chronicling the nature of a successful teaching/learning context, but also of better understanding the origins and trajectory of that context, so that we might build others like it (Gutiérrez 2002a).

Again an important aspect of this math teacher community moved beyond mere access and achievement (in terms of how many students made it to calculus) to include issues of identity and power. Identity issues included language and culture, but in complex ways (Gutiérrez 2002b, 2008). Teachers did not rely upon Mayan mathematics or some pre-scripted contexts for Latina/o students such as tortillas instead of bread. Rather, they developed a deep understanding of their students (e.g. who uses Spanish when and with whom, who prefers graphing calculators to paper-and-pencil forms, who is a leader in the school, etc.) and used that knowledge to create working groups and an atmosphere where students felt comfortable using Spanish or code-switching (regardless of their English proficiency levels) and negotiating that practice with non-Spanish speakers. Like the window/mirror analogy, teachers wanted to build upon the resources that students already possessed (Moschkovich 2007a), but they also saw the importance of students communicating their arguments in English – a language for which they would be held accountable on standardized tests. This meant sometimes students helped each other present their work. Identity issues also came through in the potlucks that teachers hosted. Students celebrated their mathematical successes with family members and invited speakers, in the midst of home-cooked food.

While more attention has been brought to the kinds of clear-cut curricular interventions that can give power to students (e.g. using mathematics to explore whether there is discrimination in the ways banks loan money), the issue of power in this teaching context related more to student voice/ownership in the classroom and to an understanding of the ways mathematics and power are related in society. At the time of the study, the calculus students showed outward signs of agency (e.g. developing T-shirts that claimed the calculus space, creating a second 'honors' assembly because their efforts had not been acknowledged in the larger school's gathering, creating a body of calculus representatives that provided feedback on teaching to their instructors).

However, the true nature of power became more prominent a year later when I had the opportunity to follow eight of the graduates into their college years at the University of Illinois, the flagship university of the state. Having moved from their neighborhood communities where most of their interactions involved other brown-skinned, mainly working-class people, the university setting presented a new space where they were often challenged to prove themselves in terms of intellect and their right to be present. Whether it was deficit-oriented professors or white and/or middle class students with negative stereotypes of urban schools, the high school graduates argued that just bringing out 'the calculus card' was enough to change the power dynamic. That is, they understood and were able to draw upon the social capital conferred on them by having participated in a calculus program in high school.

2.3 Twenty-Three Teacher Candidates

Having learned the importance of 'community' and 'stance' in the work of effective mathematics teachers, I shifted my focus to teacher education. More specifically, I studied 23 teacher candidates who remained as a cohort for 2 years as they moved through our certification program. I wanted to know how one might develop in individuals the knowledge and disposition to teach high-quality mathematics to urban students. The context of the program in which I work is primarily white, middle-class females, strong in mathematics (mainly procedural knowledge) with little exposure to or solidarity with marginalized students. While (re-)learning mathematics in ways they were not taught is important (Ball 1988), the more formidable struggle is to get teacher candidates to recognize that not all students are like them. Part of that challenge lies in getting them to acknowledge and build upon students' frames of mind.

I was frustrated with the limitations of readings and cases studies and was committed to the idea that 'learning is becoming' (Wenger 1998). As such, I was most interested in engaging my pre-service teachers in a community of practice like the successful teacher communities I had studied. I had already spent 2 years working on a similar project with a local teacher on a professional development grant. While she was committed to the highest levels of professionalism, and engaged my pre-service teachers in a kind of community of practice, she did not hold a 'stance' on teaching that placed her mainly African American and working-class students and their needs first. She received her national board certification during the final year in which we worked together: however, in my eyes, she was only minimally successful along the access dimension of equity and unsuccessful along the other three dimensions. (See Reed and Oppong 2005, for similar results on national-board-certified teachers.) At best, the pre-service teachers in that project were able to identify beliefs and practices they would not replicate. At worst, our partnership further engrained already-held stereotypes of working-class students and students of color.

For the new project, I chose a teacher who had won awards for his teaching of calculus at the college level, who chose to teach in an alternative high school serving students who had been unsuccessful in other schools, who put his Latina/o and African American students and their needs first and who chose to teach an NSF-supported mathematics curriculum. Although only in his first year of teaching, he offered greater opportunities for modeling the kind of equity practice for which I was looking. As such, we engaged in a 1-year partnership with him and his students.

The partnership project had several components that attempted to engage pre-service teachers in the kinds of practices that effective teachers of marginalized students do on a regular basis (Gutiérrez 2004). The university students were asked to: email a high school student on a weekly basis about things other than just math class, do mathematical problems that the high school students were doing, view videotape of the high school students doing the same problems, think about those math problems from the point of view of the student and the teacher, debrief with the partner teacher the events on the classroom video, prepare lesson plans for the high school classes and host a field-trip to the University of Illinois where the high school students would be given a chance to understand college life.

The success of the teaching context with which we partnered lay in a high percentage of students engaged on a daily basis with Interactive Mathematics Program (IMP) activities, focusing on conceptual understanding. Although no standardized achievement data was available, most of the students received solid grades in the two courses with which we partnered: algebra and data analysis/probability. Because the high school students had left their previous schools for reasons of child-care, gang involvement or lack of support, their commitment to the math classroom here signaled a certain level of achievement. Like many of the math departments I have studied, effective teaching in this context involved a heavy reliance on cooperative learning, emphasis on National Council of Teachers of Mathematics (NCTM) process standards, students working in Spanish *and* English, regular use of graphing calculators, student presentations of their work, a rigorous curriculum and supplemental activities that were relevant to students' lives.

Because I was interested in developing in pre-service teachers their knowledge and disposition to teach high-quality mathematics, the partnership project aimed to get them to experience mathematics in ways that reflected the goals of the NCTM Principles and Standards (NCTM 2000). The students had read and discussed the principles and standards, some agreeing more than others that it represented guidelines for a high-quality mathematics curriculum. Now, they were being given an opportunity to do activities from a real text, to see high school students doing those very activities, to video-conference with their teacher about his approach and its consequences and, lastly, to decide for themselves whether this represented a high-quality mathematics curriculum.

At the beginning of the project, most of the teacher candidates were impressed with the manner in which the Interactive Mathematics Program engaged students in working with concepts and not just procedures and offered them opportunities to connect their mathematical understanding with other topics or the real world. On first pass, and with themselves as the reference frame, the pre-service teachers saw the curriculum as high-quality mathematics. However, when asked to reflect on the students with whom we were partnering and how this curriculum might be appropriate, they were less sure, pointing out that the IMP curriculum assumed a certain level of proficiency in basic skills (something they did not feel the students had) and offered few opportunities to practice the ideas learned. They were also concerned that students could get lost in the heavily reading-based text. Their perspectives failed to engage equity issues fully, reflecting somewhat of a deficit framing on the students of color who were our partners.

When the pre-service teachers had the opportunity to view video of the high school students doing the activity, they were happily surprised to find that the high school students were engaged in reasoning and problem solving, making conjectures and defending their arguments. At the end of this session, many of them changed their minds and saw the power of a curriculum that focused on concepts and that required students to collect their own data. They saw the manner in which the teacher framed his questions to draw out his students' thinking and fostered student–student interactions. In fact, some argued that this kind of curriculum and teaching was at the heart of addressing equity issues in school, because so many inner city kids were usually only asked to memorize procedures or prepare for standardized tests. From our equity lens, they were now able to see issues of access (to a high-quality curriculum and teacher) and some achievement (ability to make conjectures and defend arguments).

As the year progressed, we became more and more familiar with the classrooms with which we partnered and developed a more natural feel to the debriefings of lessons with the teacher. As we gained trust and shared common rituals, the pre-service teachers were able to pose more nuanced questions and our partner teacher was able to be more vulnerable with us. In community, we sought to understand better the students' needs and how best to support them to develop their mathematical proficiency.

At one point in the year, the teacher was recounting a situation that had happened in his class and he wondered aloud if he had done the right thing. He explained that one day in class he was lamenting how the context of the problems in IMP failed to address the lived realities of his students. As the conversation was pursued between him and his students, the idea of white textbook writers arose and what might this curriculum look like if the high school students had written it instead. At the time of the conversation, he was aiming to show solidarity with students, recognizing that their identities and lived realities were important. However, upon later reflection, he wondered whether bringing up the subject and showing his disappointment with the curriculum would now make it hard for students to want to do the activities in class on Monday. Should he even have said anything to his students?

This question stimulated much discussion among us and raised the issue of whether 'high-quality curriculum' could be considered in universalistic terms. Several of the students saw the importance of recognizing the bias in curricular materials with one's students. In the end, the teacher ended up creating a separate project on the probability of 'seeing oneself' in a variety of magazines in his data analysis and probability class.

For the first time, my pre-service teachers were starting to see issues of identity and power. They realized that they had not considered whose perspective was privileged in designing curriculum and that teaching involves making these kinds of in-the-moment decisions that can create or break down solidarity and trust with students. More than just having access to an NSF-supported curriculum that challenged them to reason, problem solve, communicate, make connections and represent mathematics to each other and to their teacher or being able to use Spanish in class, they saw that students also should be given opportunities to see themselves in the curriculum or analyze the world around them. At the end of the project, what was less clear for the pre-service teachers was the extent to which students were being prepared for standardized tests they might encounter in their lives. Their struggle highlights the tensions between dominant and critical mathematics.

What this study revealed to me was the importance of successful contexts, not just for students in public schools, but also for developing teachers. That is, a key feature of the context of learning that differed between the teachers with which we partnered was being able to see real outcomes for students in ways that began to address their identities and power in society, especially with a teacher who held a stance of solidarity with his students. In other words, my pre-service teachers were only able to abstract ideas and strategies from things they witnessed or participated in. This feature of success further highlights how all communities of practice are not equal (Gutiérrez 2005).

2.4 The 'Achievement Gap'

I would like to end with some research I have been doing on the achievement gap (Gutiérrez 2008; Lubienski and Gutiérrez 2008), because it helps illuminate how our attention has been diverted from broader equity considerations. 'Gap gazing', along with more general 'gaps analyses', provides an example of the kind of work that is currently embraced in the U.S. mathematics education research community as a way to address equity. By 'gap gazing', I mean research that documents the gap in mathematics achievement between rich and poor students and between primarily brown/black students and white students, while offering little in the way of intervention. In its most simplistic form, this approach points out there is a problem but fails to offer a solution. Even researchers who conduct gaps analyses with the purpose of closing 'the gap' fail to recognize that it is the analytic lens itself that is the problem, not just the absence of a proposed solution. Though I see many more problems with using the achievement gap as an analytic lens (see Gutiérrez 2010), two concerns that are pertinent here are: (a) it abstracts data from contexts; (b) it ignores the many successful contexts serving marginalized students that have been documented in the literature.

Most of the research conducted on the achievement gap involves large-scale data sets. In these data sets, there is little room for attending to local dynamics, as the purpose is to define generalizable trends. However, by failing to attend to contexts, a 'gaps' focus renders policies as 'one size fits all', even though we know that teaching and learning are not universalistic (Ladson-Billings 1995a, b). That is, such analyses fail to attend to the meanings that students and teachers ascribe to practices and resources that are at their disposal. In addition, the most significant variables shown to close the achievement gap do so only minimally and do not involve schooling contexts (Lee 2002; Hedges and Nowell 1999). Rather, they focus on income or family background, something over which few mathematics educators or researchers have any control. Partly because gaps analyses provide little understanding of successful learning contexts beyond a few static variables, and also because they rely on correlations, they are almost useless in helping us understand either the dynamic relations between these variables or how to develop more such effective learning environments. Moreover, without the larger sociopolitical frame, achievement gap analyses perpetuate the notion that the problem of low achievement in mathematics is a technical one. In other words, if only we knew better how to develop teacher knowledge or teach students of color, we could close the gap.

The fact is, we know quite a bit about what is successful in terms of teaching marginalized students mathematics. For example, we know that effective teachers of diverse students (especially teachers of Latinos and African Americans) come to know their students in meaningful ways (e.g. do not rely upon stereotypes, are able to relate to their students in ways that attend to their mathematical and personal needs, build upon their cultural/linguistic resources), scaffold instruction onto students' previous learning experiences without watering down the curriculum, create classroom environments that have the feel of 'family' (including a heavy reliance on group work), believe all students can learn advanced mathematics and draw upon a deep and profound understanding of mathematics when choosing tasks (NRC 2004). Programs such as QUASAR, MESA, Project SEED and, to some extent, the Algebra Project have had success with students who are not being reached by traditional schooling practices (Hilliard 2003). What often is lacking is not the knowledge, but the public 'will' to support or develop more successful learning contexts such as these.

Perhaps more importantly, an achievement gap focus fails our definition of equity, as it attends only to the dominant mathematics that comprises the access and achievement dimensions. Equity problems among students are complex; no one variable is the lever. Therefore, although the policy arena may pressure us to keep things simple, the designs of our studies and ultimately our solutions must mirror that far greater complexity.

3 Future Research

My point in highlighting some of the research I have conducted on successful learning environments for marginalized students is not to say that these environments are generalizable to all students. Rather, it is to suggest that only in deeper exploration of these environments can we begin to understand the meanings that emerge for teachers and students. Moreover, my emphasis on the potentially dangerous consequences of using an achievement gap focus is not to suggest that everyone must conduct the same research or even that marginalized students do not benefit, at times, from studies that document inequities. However, in an era of randomized trials and experimental designs, I argue there is a great need to reclaim a space for studies that focus on learning in context, *especially* if we are committed to a definition of equity that moves beyond mere access and achievement.

More specifically, I encourage the mathematics education community to conduct less research that documents an achievement gap, identifies causes of that gap and/or focuses on single variables to predict student success. Instead, we need more research on effective/successful teaching and learning environments for black, Latina/o, First Nations, English language learner and working-class students. More rich descriptions of these contexts, including their origins of development, are necessary if we are to engage fully with a diverse society. Indeed, we need to learn more about and build upon effective models that already exist. From these studies more intervention work is possible.

A cursory read of this chapter could leave one wondering if I am calling for the erasure of all large-scale quantitative research or research on populations other than marginalized students. I am not. However, I am challenging us to consider the ways in which the contexts we study come through in our work and how that may relate to our stated goals of equity.

4 Postscript (2011)

Three years have passed since writing the foregoing and my views have evolved. Although I still find the four dimensions of equity (access, achievement, identity and power) important in my work, I now place a greater emphasis on the tensions that are created when one tries to balance the critical and dominant axes in mathematics education (Gutiérrez 2009). In working with pre-service teachers, I find it particularly useful to recognize multiple perspectives and to maintain these tensions in one's practice as opposed to looking for quick solutions. It is from within these tensions and uncertainties that educators and learners birth new knowledge.

Moreover, I find it useful to highlight the sociopolitical dimensions of mathematics education: how research in whitestream mathematics education differs from that of researchers who focus on equity issues; how mathematics is defined; how it relates to schooling as an enterprise; what this means for the identities that are available to individuals and groups (Gutiérrez 2010). I now put greater emphasis in my research onto *education* as opposed to just *schooling* (Gutiérrez and Dixon-Román 2011). The examples of research in this chapter all occured within schools, suggesting such institutions are the only focus to which researchers should attend. Yet we are educated through all aspects of our lives, including through media, churches, families, on street corners, by product labels, etc. In this sense, mathematics has an influence not just on our educational experiences and identities within classrooms and schools, but mathematics also formats our lives by providing a lens onto our worlds. It is only when we are able to question the practices that occur within schooling, as well as those that operate beyond it, and in conducting mathematics education research that we can really begin to address a broader sense of equity (and transformation) in society.

Chapter 3
Exploring Scholastic Mortality Among Working-Class and Indigenous Students

Robyn Jorgensen

In Australia, the national psyche is one of egalitarianism, a 'fair go' for everyone and where mateship is fundamental. Historically seeking to create a nation different from the 'mother country', Australians generally subscribe to a view that everyone is equal and that it is a nation not divided along class lines, while nonetheless acknowledging the large gaps in health and well-being between Indigenous Australians[1] and non-Indigenous Australians. But these views of a relatively egalitarian nation, one where people can succeed with hard work, have been shattered by the results of international testing schemes.

As a nation, Australia performed reasonably well on tests such as TIMSS and PISA, ahead of many similar countries. Of concern to educators, however, is the massive tail in performance. McGaw (2004) has been unrelenting in his challenge to the egalitarian myth through the outcomes of these studies. The myth of equity has been shattered by the finding that Australia may have performed well on PISA, but it has the longest tail of all OECD countries, making it the poorest performing in terms of equity.

This chapter explores the ways in which the discourses of mathematics contribute to this tail for working-class and Indigenous Australians. These two groups, along with students who live in rural and remote settings, are most likely to be in the 'tail' of the performance measures. Most often there are intersections between these categories. For example, remote Indigenous learners are the most likely to be the poorest on performance measures. The intersection of these categories exacerbates the disadvantage experienced in schooling. While the chapter focuses on the role of

[1] The term 'Indigenous' is used in this chapter to refer to Aboriginal and Torres Strait Islander People. It is recognised that the term is problematic, as it fails to acknowledge the great diversity among the first nation peoples of Australia. It is used here as a marker to highlight the diversity of the original inhabitants of Australia.

R. Jorgensen (✉)
Griffith Institute for Educational Research, Griffith University, Gold Coast, QLD, Australia
e-mail: r.jorgensen@griffith.edu.au

B. Herbel-Eisenmann et al. (eds.), *Equity in Discourse for Mathematics Education: Theories, Practices, and Policies*, Mathematics Education Library 55,
DOI 10.1007/978-94-007-2813-4_3,

English as the medium of instruction, the propositions developed in this chapter will have applications to other English-speaking contexts.

In terms of equity in mathematics education, I take a relatively structuralist position to identify the practices within the field that reinforce the dominance of the powerful groups while ensuring the subjugation of less powerful groups. This is not to present a deterministic approach, but rather to enable discussion around marginalising practices with the explicit intent to redress such practices. Rendering visible those practices which support the status quo enables challenges to them to be made and thus supports greater access to mathematics. Simultaneously, it calls to task the knowledge structures and social practices implicated in the construction of inequalities in mathematics education.

I frame my discussion using Pierre Bourdieu's theoretical constructs. In particular, his concept of miscommunication is most appropriate for helping to understand the ways in which working-class and Indigenous learners are excluded from participating in mathematics education. Bourdieu, illustrated through his comprehensive writing presented in *Academic Discourse* (Bourdieu et al. 1994a), seeks to understand how educational institutions (in this case, the academy) create opportunities (or not) for success. I have adopted and adapted this work to the context of school mathematics for this chapter.

Overall, this chapter is framed by Bourdieu's proposition in which he sums up the processes through which hegemony is realised through the discourses of academic institutions. Although Bourdieu does not specifically address school mathematics, his fundamental premises can be appropriated for this context. His work is particularly powerful in coming to understand the marginalisation of social groups through the discourses used within school mathematics.

> To fully understand how students from different social backgrounds relate to the world of culture and more precisely, to the institution of schooling, we need to recapture the logic through which the conversion of social heritage into scholastic heritage operates in different class situations. (Bourdieu et al. 1994b, p. 53)

The notion of social heritage thus becomes a central variable in coming to understand differential success in school mathematics. Using a Bourdieuian framework, the lack of success for some social groups becomes a non-random event, a product of institutionalised practices of which participants may be totally ignorant. School mathematics represents a particular and powerful example of how social heritage can convert to academic success. Language is an integral part of the social heritage that is brought into school mathematics and becomes reified as some form of innate ability that facilitates, or not, success in coming to learn the disciplinary knowledge within the field of school mathematics.

In this way, success in school mathematics has less to do with innate ability and more to do with the synergistic relationships between the culture of school mathematics and that which the learner brings to the school context. The greater the synergy between the linguistic habitus of the student and school mathematics, the greater the probability of success. In Bourdieuian terms, the habitus thus becomes a form of capital that can be exchanged within school mathematics for forms of recognition and validation that convert to symbolic forms of power. Such manifestations of this conversion can be seen in grades, awards, scholarships and other forms of accolade.

The difference in language brought to the school context has been evident in many studies. In particular, the work of Walkerdine and Lucey (1989) has shown how the relationships of signification between mother and child are very different depending on the social class of the family. For example, in their study they found that middle-class mothers were more likely to use the signifiers 'more' and 'less' in their interactions with their children. Conversely, working-class mothers were more likely only to use the signifier "more" in their interactions. When considering much of the early years of schooling, mathematical ideas are often taught through comparisons – 'which number is two more than x?', 'what number is 5 less than 7?', 'which is more 3 or 8?', and so on. These same comparisons apply to other areas of the applied curriculum (e.g. measurement) where students need to undertake comparisons. Within this framework, it becomes a substantive issue if some students bring to the school situation a working knowledge of the twin concepts of more and less in relation to their working-class and Indigenous peers, whose restricted vocabulary limits access to certain mathematical concepts.

Similarly, in her seminal work, Heath (1983) explored interactional styles within families, comparing them with those used in schools. She found that working-class families tended to be more staccato in their ways in communicating, with a strong propensity for families to rely on declarative statements when interacting. This is different from the discursive practices of schooling that tend to be more rhetorical in form. This difference in questioning styles creates dissonance for working-class students as they come to learn (or be excluded from) the participatory practices of engagement in classrooms. Similar differences occur in the interactional patterns of many Indigenous learners as they enter the school context. These will be expanded upon in a later section of this chapter.

These studies reinforce the original work of Bernstein (1982) who proposed that the school code was that of the middle class. He differentiated between the linguistic codes used by the middle class and the working class. While his nomenclature of 'elaborated' and 'restricted' codes created controversy, by suggesting that the working-class 'restricted' code was in some way deficient, this was not his intention. Rather, he was seeking to highlight that the code of schooling was strongly aligned with that of the middle class.

Student and teacher behaviour is complicit in the stratified outcomes of learning school mathematics. Where there is little or no recognition of the linguistic codes that learners bring to school mathematics, there is greater probability of miscommunication and hence scholastic mortality. For example, where a working-class child enters the classroom with the register of the working class, the child may speak English, but it is of a different form from the English register of the classroom. The teacher may assume that since it is English that is spoken by the child, then he/she understands the discursive interactions and is able to participate fully in such interactions. In reality, this is far from the truth and much of what happens in that interaction may be misconstrued or misunderstood by the student. This process of miscommunication becomes a subtle form of exclusion of which both the child and teacher may be totally ignorant. The miscommunication is transformed into scholastic mortality whereby the child is excluded from participating and learning particular academic and/or social knowledge needed for being seen as 'successful' in the school context. This difference between the home and school languages is summed up by Bourdieu thus: "the more distant the social group from scholastic language, the higher the rate of scholastic mortality" (Bourdieu et al. 1994b, p. 41).

In the preceding paragraphs, I have sought to highlight the ways in which the habitus, of which language is an integral part, is implicated in the success or not of students as they come to learn school mathematics. For some students, the chances of learning school mathematics are limited by their linguistic, social and cultural habitus. This is not to propose a deficit model of success, but rather a difference model – a difference between the practices of schooling and the habitus of the learner. Those most likely to succeed in the discipline are those whose habitus is strongly aligned with the objective structuring practices of the field – in this case, school mathematics.

This process works both at an ideological and at a social level. By being ignorant of the mismatch between the habitus of the learner and the regulative practices of the field, teachers may inadvertently exclude students on the basis of the social and cultural backgrounds as represented in and through the learner's habitus. Being aware of the subtle ways in which incongruencies between the field and habitus create disjunctions for some students, teachers who are aware of the processes of scholastic mortality may be better positioned to address and change the exclusionary practices within mathematics (and other curriculum areas).

1 Teaching Working-Class and Indigenous Students: An Act of Symbolic Violence?

The current modes of teaching mathematics to working-class and Indigenous students can be seen to be an act of symbolic violence when using a Bourdieuian framework. Such an approach implies that learners and teachers are complicit in the reproduction of dominant forms of knowledge to the detriment of marginalised groups, in this case, working-class and Indigenous students. The processes implicit in the act of teaching allow dominant forms of knowledge and knowing to retain their hegemonic role through the exclusion of working-class and Indigenous forms of knowledge and ways of knowing. For Bourdieu, this is summarised as follows:

> the theory of symbolic violence rests on a theory of belief or, more precisely, on a theory of the production of belief, of the work of socialization necessary to produce agents endowed with the schemes of perception and appreciation that will permit them to perceive and obey the injunctions inscribed in a situation or discourse. (1998, p. 103)

For example, concepts such as ratio and proportion are not taught in Australian schools until the upper primary years as they are considered difficult concepts. The examples used when teaching such concepts are often the 'best buys' of products; that is, value for money. Yet, as the research on everyday practices such as shopping (Lave et al. 1984) has shown, other considerations are used when purchasing. For many working-class families, activities outside the home have a function. Activities are rarely undertaken for a recreational or solely educational purpose, so that activities such as fishing are common. In this case, there is considerable work in ratio, since the size of the hook is commensurate with the type of fish to be caught. Examples from working-class practices are rarely used in mathematical texts, yet they are rich mathematically and experienced quite early in the family socialisation process. Rather, symbolic violence occurs in the representation of bourgeois activities in mathematical texts so as to normalise middle-class activities while excluding working-class and/or Indigenous activities. It is as if shopping for value is a much more valuable practice than fishing.

But symbolic violence does not occur without considerable preliminary work to be undertaken. Knowledge and pedagogy are twin considerations when taking into account how symbolic violence is enacted in schooling. In the case of mathematics, there have been centuries of repression of forms of knowledge such as those of the Islamic and Asian contributions to mathematics (Joseph 1991), so that legitimate knowledge of the curriculum is perceived to be solely that of a western orientation and where other forms of knowledge are absent from the knowledge base. Through this process, a reinforcing of Western ways of knowing have led to them becoming the taken-for-granted forms of knowledge.

Similarly, the ways of teaching have become accepted as the dominant modes of pedagogy. The pedagogy of schooling supports middle-class and western modes of learning. For example, the individualistic and competitive styles of classroom organisation in mathematics reinforce middle-class values. Considerable work in the reform of school mathematics has been undertaken, but as Gutiérrez (1998) along with many other mathematics educators assert, the change process in mathematics is very difficult. This reflects acceptance of the taken-for-granted ways of teaching mathematics that have become resistant to change despite considerable research to show that such practices do not meet with success. The change classrooms in Boaler's work highlight how changes in pedagogy open up learning opportunities for disadvantaged students, resulting in considerable successes for those students (Boaler 1997, 2008).

> For the symbolic act to exert, without a visible expenditure of energy, this sort of magical efficacy, it is necessary for prior work – often invisible, and in any case forgotten or repressed – to have produced, among those who submit to the action of imposition or injunction, the dispositions necessary for them to feel they have obeyed without even posing the question of obedience. (Bourdieu 1998, pp. 102–103)

Thus, much of the teaching of school mathematics can be seen as an act of symbolic violence when undertaken in Indigenous and/or working-class contexts. This is particularly the case where cultural forms of knowing are not an integral part of the curriculum.

Where the complicity between beliefs and practice becomes apparent is when teachers blame lack of success on factors such as attendance, health, transience, family expectations, and so on and then offer an impoverished curriculum for students. But this need not be the case. In the area of literacy, Gray (1999) has argued that such factors should not be seen as an excuse for poor performance of Indigenous students. He contends that with appropriate scaffolding, teachers can offer a very rich literacy program for Indigenous students where success is more than possible.

Likewise, there is every potential for all students to come to learn mathematics, provided there is adequate provision made to scaffold students' learning to enable a bridging between the two cultures. However, as Boaler's (2008) comprehensive work has shown, learning environments are critical to the success of these students. Transplanting a traditional pedagogy into a classroom where there are considerably disadvantaged students offers little chance for success. Rather, as her detailed study of 'Railside School' has shown, by radically changing pedagogy to meet the cultural dispositions of students while offering opportunities for rich mathematical learning, students can succeed in school mathematics. In this case, it included the use of home languages to allow students to negotiate meaning in their small groups, despite a one-language policy in that state.

Much of the current teaching of mathematics involves transmission of knowledge that is inherently that of the dominant classes. There has been a strong criticism of the hegemonic forms of knowledge that are embedded in school mathematics. There is a vast literature in this area that can be accessed through the proceedings of conferences such as *Mathematics Education and Society*. Similarly, the ways teaching approaches are organised can similarly create difficulties for learners when their cultural knowledge and ways of knowing do not form part of the repertoire of teaching practices. The need for an approach that is culturally inclusive has been recognised by some educators:

> Teaching methodologies that recognize and build upon the pupil's cultural heritage and the specific ways in which children are taught to process information can play a critical role in addressing this concern. Inclusive instructional strategies that recognize and embrace this cultural dimension can help teachers ensure that all students in mathematics classrooms become successful learners. (Varghese 2009, p.14)

Indeed, many of the intervention programs used in mathematics education reform in Australia maintain the status quo of the field by taking for granted methods and knowledge forms of school mathematics and then seeking to impose them on working-class and Indigenous students. For example, the work of Siemon (2008) is an example of this approach. In her work, she promotes the key ideas in mathematics that are part of Western ways of knowing. These ideas are then taught to Indigenous students in ways that fail to problematise the knowledge as if it is open to all. In her work with Indigenous students in Western Australia, Willis (2000) challenges such an approach where the fundamentals of mathematics curriculum design are seen as the 'right' way to organise curriculum. In her work, Willis found anomalies in the ways in which some Indigenous students were able to subitise before they could count. Her work throws out a challenge to approaches that take as unproblematic representations of school mathematics reified in curriculum documents.

There have been a number of ethnomathematical studies that have sought to 'unfreeze' the hidden mathematics in activities undertaken by under-represented groups. For example, two cases can be used to illustrate this. In her comprehensive work of Indigenous communities in the Northern Territory, Harris (1991) reported the mathematical concepts in Indigenous art and, in doing so, made claims that Indigenous people had understandings of a number of mathematical ideas (such as symmetry, concentric circles, parallelism). Similarly, in their work in Indigenous communities in Queensland, researchers claim to identify the mathematics undertaken by students in card games (Baturo et al. 2004) that are common to many Indigenous communities. Collectively, such approaches take the Indigenous activity and search for the Western mathematics within that activity. Dowling (1991) has convincingly argued that such approaches reinforce the (high) status of Western mathematics while subjugating the Indigenous activity to Western mathematics. These approaches seek to justify that Indigenous people have mathematical understandings and hence are able to do mathematics. However, these studies prioritise western mathematics while failing to understand the activity from the perspective of the participants.

2 Both-Ways Education: Challenging Symbolic Violence

A very different approach is to create synergies between the different cultures so as to minimise symbolic violence. Ezeife (2002) illustrated the power of building bridges between the culture of the students and that of school mathematics. An example of this type of approach is that of Watson (1987) who worked with the Yolgnu people in Arnhem Land in Northern Territory to develop a both-ways education program. Working with the Yolgnu people, they developed a mathematics program that recognised both Western and Indigenous ways of knowing as legitimate. In this work, for example, the ways of understanding the land (mapping the landscape) were undertaken through both school mathematics and Indigenous approaches. The grid reference system used in school mathematics was significantly different from the ways in which Indigenous people 'sing and sign the land' through historical and cultural markers (Watson and Chambers 1989). Such an approach seeks to develop border crossings between one culture and another, with neither culture being foregrounded while acknowledging that each has particular strengths. Indeed, it could be argued that incorporating Indigenous knowledge into school mathematics may enrich the experience of all students.

3 Resistance Theory: Alternatives to Deficit Models

The *National Inquiry into Rural and Remote Education* (HREOC 2000) suggested that Indigenous Australians "have become alienated from the school system" (p. 58). Opting out of schooling and school mathematics by Indigenous and working-class students can be understood as an active process of resisting the imposition of the dominant culture. In part, this is due to the lack of synergy between the parents' and communities' expectations of schooling and what schools offer. When framed within Bourdieu's project, the difference between the culture of school mathematics and the culture that the students bring to schools does not arise simply as an apolitical act. Rather it is one where the misrecognition of the two cultures enables mathematics to maintain its power base and for those who fail it allows them to assume something inherent about their own ability and thus become complicit in the production of their own oppression. This, as Bourdieu noted above, is done in a way that remains hidden to those participating in the act. Within this framing, truancy that is seen to be an endemic issue among Indigenous students may be a reasonable act of resistance to the symbolic violence being enacted upon them and their families. It would seem a priority within education provision that border-crossing pedagogy and curriculum becomes an essential component of educational reform if Indigenous Australians and working-class students are not to vote with their feet.

In the remainder of this chapter, I provide specific examples from my research in working-class and Indigenous Australian classrooms to illustrate the ways in which discourses and practices work to exclude students from learning mathematics.

4 Language Structures and Marginalising Discourses

The primary socialisation of children through the familial contexts creates a particular habitus. This habitus includes language. How this linguistic habitus aligns with the discourses of school and school mathematics creates greater or lesser chance for success. As I have argued elsewhere (Zevenbergen 2001b), the linguistic habitus that students bring to mathematics classrooms enables greater or lesser chance of success. These successes need to be understood at multiple levels of discourse: it is about 'cracking the code' of mathematics (Zevenbergen 2000). This can be at the multiple levels of language from words to sentences through to larger bodies of text (Zevenbergen 2001a), as well as at that of the social context within which learning occurs. This includes the ways in which teachers attempt to create meaning-making opportunities through the pedagogic relay (Zevenbergen et al. 2004).

Scaffolding techniques and questioning are central to these analyses. Bourdieu claims that the early socialisation of learners creates a primary habitus that shapes the forms and uses of language. For those students for whom the primary linguistic habitus resonates with the discursive practices of the mathematics classroom, along with the linguistic registers through which mathematical concepts are relayed, there is greater chance of making sense of the interactions among teacher and peers. Bourdieu sums this position up thus:

> The ability to decode and to manipulate complex structures, whether logical or aesthetic, would appear to depend directly on the complexity of the structure of the language first spoken in the family environment, which always passes on some of its features to the language acquired at school. (Bourdieu et al. 1994b, p. 40)

Thus, in part, the role of education becomes one of reconstituting the linguistic habitus of marginalised learners so as to align it with the valued practices within the field. For working-class students, this requires the reconstitution into a middle-class habitus. For Indigenous students, this may require multiple reconstitutions of the habitus so it becomes Westernised, Standard Australian English and middle class. When seen in this way, there is considerable work to be undertaken for learners to succeed in mathematics.

5 Language Differences: The Case of Questioning

In this section, I use the case of questioning as an illustration of how the habitus of learners is differentially recognised within mathematics education and how this differential recognition contributes to the marginalisation of disadvantaged students in subtle and coercive ways. I do not intend to discuss the literature on questioning types; rather, I focus on the social practice of questioning *per se*. I draw heavily on Bourdieu's notion of 'the game'. He uses this notion in a highly specific way to connote the social practice. Like a game, there are rules that must be complied with, some of which may be unknown. Similarly, some people may hold more trump cards than others and are thus more likely to be winners. In Bourdieu's game, these trump cards are often the cultural capital that the player brings to the game. In the game of school mathematics, the trump cards can be seen as language of the learner; her or his exposure to number and other mathematical concepts and processes; familiarity with technologies such as computers or calculators. All such characteristics enable certain students to play the game with a stronger hand; that is, each characteristic can be considered as a trump card.

5.1 Written Questions and Contexts

In their seminal work on wide-scale national testing, Cooper and Dunne (1999) showed how there were negligible differences between middle-class and working-class students on esoteric mathematics problems. However, when the mathematics was embedded in 'realistic' problems, middle-class students performed much better than their working-class peers. These differences were not due to some inherent ability, but instead was due to the 'misrecognition' (Bourdieu 1990) of the implicit demands of the question. Middle-class students recognised the 'realistic' problems as mathematics questions and were cognisant of the need to respond in the appropriate discourse. By contrast, working-class students were more likely to perceive the problem to be realistic and respond using the discourse within which the problem was posed and hence respond with a much more practical answer, which was deemed incorrect.

When exploring problems with working-class students (Zevenbergen and Lerman 2001), the following example was provided: "There are 365 students at the sports field. If a bus can hold 50 people, how many buses are needed to transport students back to school?" The range of responses included '8'; '7 rem. 15'; '7 but some will stand up, sit 3 to a seat'; '7, some can come back with teachers, parents'; '7 and a mini bus'. As with Cooper and Dunne's study, we found that the working-class students often interpreted the question as a real one about buses and sports carnivals, whereas this mistake was less common among middle-class students. In working with teachers, this research has resulted in them making the implicit demands of the question explicit to the students. It requires the teacher to scaffold the learners to understand better the pedagogic register of mathematics classrooms.

Within a Bourdieuian framing, the practice of questioning can be seen as a game in which players are more or less predisposed to participate in the game depending on the social heritage that they bring to the game. Using Bourdieu's game metaphor, it is about the trumps that they bring to the context. Knowing how to play the game becomes critical to one's success, but the rules of the game are rarely made explicit to players:

> One does not embark on the game by a conscious act, one is born into the game, with the game; and the relation of investment *illusion*, investment, is made more total and unconditional by the fact that it is unaware of what it is. (Bourdieu 1990, p. 67; *italics in original*)

Participating in classroom interactions is made more or less easy depending on what one brings to the field – in this case, mathematics education. Those who participate easily are not likely to understand explicitly the rules of interaction, but come to the game already knowing the rules from other contexts – in this case, the home.

5.2 Classroom Interactions

Typically, a mathematics lesson is one where the flow of knowledge is sought to be achieved through the teacher controlling the lesson through the use of questions. The questions may be used to promote mathematical understanding, as well as to control the flow of the content (Lemke 1990). The questions are typically posed by the teacher who knows the answers to the what is being asked, so it becomes a game where the students are expected to respond in particular ways to what is being posed. Teachers buy into this game through their socialisation into teaching as a profession, so that it becomes part of their habitus as a professional. They often do not call into question the practice itself, but rather accept it and hence reproduce it unquestioningly. Bourdieu (1990) explains this process thus:

> The earlier a player enters the game the less he [sic] is aware of the associated learning… the greater is his ignorance of all this is tacitly granted through his investment in the field and his interest in its very existence and perpetuation and in everything that is played for in it and his unawareness of the unthought presuppositions that the game produces and endlessly reproduces, thereby reproducing the conditions of its own perpetuation. (p. 67)

5.3 Questions to Control Behaviour and Flow of Lessons

Lemke (1990) adopted the use of Sinclair and Coulthard's (1975) tripartite model of interaction around teacher questioning – Initiate, Respond, Feedback (IRF) – which occurs commonly in classrooms. In this interaction, the teacher poses a question, the student responds and the teacher then provides evaluative feedback (often explicitly) on this response. This is illustrated in the following example:

1	T:	What does area mean?
2	S:	The outside of the square
3	T:	Not quite, someone else? Tom?
4	S:	When cover the whole surface, that's area.
5	T:	That's good

Lemke proposed that this practice was used for many purposes in the classroom, including controlling both behaviour and the pace of lessons. It was minimally used for eliciting deep knowledge. In comparing three very different schools based on their class backgrounds over a year-long period (Zevenbergen 1995), I consistently found that middle- and upper-class students complied readily with this pattern of questioning interaction, thus enabling the flow of lessons. However, in working-class lessons, there was considerable chaos as the students did not comply with the IRF model. This was at all year levels. At Year 6,[2] working-class students did not comply with the expectations of the teacher's use of IRF in the mathematics lessons, as I will illustrate.

1	T:	So, if I put those together we start talking more about a shape I am talking about. It's sort of a rectangle on the sides, all the way round but you don't call it is a rectangle, because a rectangle is just the flat surface. What do you call the whole thing if that was one whole solid shape. What do you call that?
2	C:	A cube.
3	T:	He said a cube. Don't call out please.
4	C:	A rectangular rectangle.
5	T:	You're on the right track.
6	C:	A 3D rectangle.
7	T:	Three dimensions, technically I suppose you're right.
8	C:	A rectangular
9	T:	It's a rectangular something. Does anyone know what it is called?
10	C:	A parallelogram.
11	T:	Put your hand up please.
12	C:	[unclear]
13	T:	No.
14	[More calling out]	
15	T:	I guess you could have a rectangular parallelogram, but no. A rectangle is a special parallelogram.
16	C:	A rectangular oblong.
17	T:	The word we are looking for is 'prism'.

(continued)

[2] In this study, Year 6 was the final year in primary school in the state in which the study was conducted. Typically, students were about 12 years old and had experienced 7 years of formal schooling. This is common in most Australian states, except for Queensland and Western Australia where a different numbering system is used and the final year of primary school is Year 7.

(continued)

18	C:	Yeah that's what I said.
19	T:	Say the word please.
20	Cs:	Prism.
21	T:	Not like you go to jail 'prison', that's prison. Excuse me, could you return those please.
22	[calling out]	
23	T:	So one thing that we think about with rectangular prisms and that this shape on here is, excuse me…Now you can leave them down please. You need a little bit of practice at lunch because you can't stop fiddling. This shape here is drawn out on the graph, this grid here [net for a rectangular prism]. We're going to try and do the same thing. Draw the shape and then cut it out. If you look at the shape, it's made up of rectangles and squares.

What can be seen from this example is the working-class interaction is fractured and disjointed, with the teacher often digressing to address behaviour issues. At the end of turn 1, the teacher has asked a question, a student has offered a response, but has called out which the teacher admonishes. Another student offers a response in turn 4, which the teacher evaluates and subsequently uses a funnelling technique to encourage students to keep responding in the manner as offered. The funnelling technique is common as it allows teachers to evaluate partial responses in order that the partiality provides a cue to the students for eliciting responses that closer resemble the desired answer. This funnelling technique continues through the following interactions as he attempts to elicit the desired response from the students.

What can be observed over the series of interactions is that there are more transgressions of the triadic dialogue and the teacher admonishes violations to the interactions quite frequently. In some cases the rejection of responses is made explicit (turns 10–14 – "don't call out" or "put your hand up") or more subtly where responses are ignored (turns 17–19). Freebody et al. (1995) have argued that these types of interactions embody another set of unspoken instructions to students. The messages embedded within the interactions are not made explicit to students, so that they must be able to make sense of the subtext of the teacher's spoken and unspoken interactions. In these sets of interactions, the evaluations of the students' responses are either quite explicit (see turns 5, 7 and 9) or implicit in turn 3. Turn 3 provides an example of where the response is felt worthy of acceptance, but the manner in which it was proposed was not acceptable (calling out). In this line, the students need to be able to recognise the repetition of the response is to be taken as an acceptance of the response, although it is not explicitly evaluated as such. The secondary comment – "Don't call out please" – could be seen to make the initial comment redundant.

5.4 Questions to Elicit Knowledge

In considering the role of questions outside schooling, questions perform a particular function – to elicit knowledge of the unknown. This can be seen in social situations where directions to a site are needed or knowing how much an item will cost is desired. In these practices, questions are substantially different from the practices of school. Thus, the game of questioning in schooling involves a particular use of linguistic turns, of roles and power relations. Complicity in the game is essential for its effective functioning and for the creation of winners and losers.

For those who are familiar with the rules of the game, the chances for playing (and winning) are enhanced in comparison with those who do not recognise the rules. For Bourdieu (1990), those who can play the game are involved the process of "creating symbolic capital which can only be performed on the condition of the logic of the functioning of the field being misrecognised" (p. 68).

As the work of Heath (1982) has shown, practices of questioning among the classes are considerably different, where middle-class families pose rhetorical questions that serve the same purposes as the declarative statements posed by working-class families. This is obvious in the comparison of "Would you like to do the dishes?" with "Go and do the dishes" respectively.

In sharing this approach with teachers working in working-class schools, this distinction has been a revelation for many of them. Furthermore, the observations have been consistently reinforced across many schools and districts. In many instances, teachers have also indicated that the use of "please" at the end of a sentence can create a sense of an option for many working-class students. For example, the use of the rhetorical question "Would you like to get out your maths books?" may meet with little opposition in middle-class settings, as the students are complicit in the game and realise that the question is a statement without options and the task is to remove mathematics books from the desks and begin to commence working. These commands are not explicitly made, but the students nonetheless recognise the demand implicit in the question. In contrast, in working-class classrooms, teachers report that it is not advisable to pose this question, as it is likely to be seen as a choice and that many teachers prefer to make the imperative request "Please get out your maths books". But, in some cases, even the use of 'please' can be construed as signalling an option. As such, the format of questions becomes a critical aspect of classroom interactions among classed contexts.

Within Indigenous communities, the purpose of questioning is different from that of the classroom. In making the following statement, I recognise that there is a wide diversity among Indigenous people in Australia and do not want to present the following as being common across all of these cultures. In work undertaken in some remote areas of Australia, the support people who have worked in these contexts for some extended time have reported that Indigenous students find it odd that teachers pose questions when they know the answers. For these students, the purpose of the question is to find out something which is unknown, so the posing of questions to which the answer is known seems very strange.

Similarly, in some Central Australian cultures, elders judge the learning of their children by the questions they pose. In explaining stories to young children, the child is then expected to ask questions of the elder/s about the story. When the elder has finished her/his storytelling, the child may have some questions. The quality of the question provides the elder with an indication of how well the story has been understood. This purpose of questioning is reversed to that of schooling where the teacher judges the learning of the students by the responses offered to the teacher-posed questions.

6 Teacher Judgement and Success: The Curse of Ability in Mathematics Education

Perhaps the most powerful myth in mathematics education is that of 'ability'. It permeates the field so as to render the successes and failures of students as being due to some inherent aspect of the learner. In so doing, it denies the ways in which the pedagogical tools used in this field are implicated in creating these outcomes. What I have sought to undertake in this chapter is a challenge to this myth. Using Bourdieu's theoretical project to help understand the role of language practices in mathematics education in the reification of social and cultural differences, I have argued that aspects of mathematics education can be seen as a game that confers status unequally across students based on particular characteristics. Typically, such characteristics resonate with the social and/or cultural backgrounds that students bring to school. Such characteristics are embodied in the habitus, which provides a lens for seeing and acting in the social sphere. For this chapter, I have particularly focused on the role of language and discourse in the reification of differences in mathematics education.

> Of all the cultural obstacles, those which arise from the language spoken within the setting of the family are unquestionably the most serious and the most insidious. For, especially during the first years of school, comprehension and manipulation of language are the first points of teacher judgement. (Bourdieu et al. 1994b, p. 40)

What I have sought to argue is that the unequal outcomes that are so evident in the field of mathematics education are not due to some innate ability, but to the practices within the field that value particular dispositions that learners bring to mathematics classrooms. For working-class and/or Indigenous students, the probability for success in school mathematics is hindered by their familial habitus, since it is often different from that which is valued in the field. In order to be successful, considerable work needs to be undertaken by the teacher to enable the reconstitution of these learners' habitus, so as to enable them access to school mathematics. Such an approach, however, does not call into question the forms of knowledge embedded within the field and helps to preserve the status quo.

I would contend that the work of educators is two-fold if working-class and Indigenous learners are to be more successful in school mathematics. First, there needs to be some questioning about 'mathematics' and whose culture it represents. If there is some consensus that it is important for the field to remain as it is, then the task becomes one of enabling greater access to the field. This requires changes to practices to render them more explicit for those students who do not have the cultural heritage that is valued implicitly by the field. However, there may be some scope for calling into question the field itself and ask whose knowledge is represented.

When there is an absence of working-class and/or Indigenous forms of knowledge, the curriculum may be impoverished. As we have seen with gender reforms in mathematics education – where altered practices that are more 'girl-friendly' (such as in written assessment or group work) have been implemented – there have been greater successes by girls once the masculine hegemony of mathematics was challenged.

I find the use of Bourdieu's framework powerful because it enables a shift of attention from the individual to the field and, by implication, to the practices within that field, in order to understand the ways in which particular groups are marginalised in their study of school mathematics. This framing is potent in understanding better issues of both equity and access, since it highlights the subtle and coercive ways in which the practices of the field can exclude learners. But it also creates a path forward, so that practices can be challenged in order to enable greater access by those who are traditionally excluded. The miscommunication that is endemic in the field can be reversed, so that scholastic mortality can be reduced for working-class and Indigenous learners. Part of this process is in making the game explicit to those who would otherwise not be able to engage with it. Symbolic violence can be reduced when educators become aware of the exclusionary practices that are integral to but silent within school mathematics.

Chapter 4
Mathematics Learning in Groups: Analysing Equity Within an Activity Structure

Indigo Esmonde

Many mathematics teachers have adopted cooperative group work as a daily classroom practice, along with NCTM standards-based mathematics curricula and pedagogies based on constructivist views of learning (Antil et al. 1998). Proponents of cooperative learning argue that working together provides students with more opportunities to talk about mathematics, to learn from others and to learn through teaching (see, for example, Brown and Palincsar 1989; Farivar and Webb 1994; Nattiv 1994; Stevens and Slavin 1995). Still, when not implemented carefully, group work may exacerbate equity issues in the classroom by supporting students who are already successful, while leaving less successful students behind (Cohen et al. 1999; Weissglass 2000).

The equitable implementation of cooperative learning in mathematics classrooms depends not just on what teachers do; students' learning depends on how they interact with one another. A single cooperative activity structure may be taken up in multiple ways in the classroom; issues of equity are complex enough that some aspects of an activity might support equity, while other aspects detract from it. Therefore, in order to understand better issues of equity, we must examine this uptake process – within particular activities, how do students interact and what are the consequences for their learning? To this end, I present an analysis of a single type of activity in a high school mathematics class: preparing to give a presentation to the class. I study how groups took up this activity and how their uptake affected the distribution of opportunities to learn in the group.

In this chapter, I will not directly address issues of race, class and gender, although these terms are central to debates and discussions of equity in schools, and in society more broadly. Rather than beginning with definitions of equity that draw on these terms, I begin with definitions that highlight participation in classroom practices, focusing on how people choose to participate, how they shape the partici-

I. Esmonde (✉)
OISE, University of Toronto, Toronto, ON, Canada
e-mail: indigo.esmonde@utoronto.ca

B. Herbel-Eisenmann et al. (eds.), *Equity in Discourse for Mathematics Education: Theories, Practices, and Policies*, Mathematics Education Library 55,
DOI 10.1007/978-94-007-2813-4_4, © Springer Science+Business Media B.V. 2012

pation of others and how both are constrained by activity structures within particular practices. Although positioning in relation to racialized, gendered and socioeconomic categories may often be associated with positioning in relation to mathematical competence, students and teachers do not always make these forms of positioning visible in their interactions. Because positioning in terms of mathematical competence *is* often visible in classroom interaction, this type of positioning formed the basis for the analysis in this chapter.

1 Equity and Opportunities to Learn in Mathematical Group Work

In this chapter, my goal is to explore the conditions that support equity by fostering adequate opportunities to learn for all students. Although equity has not been a central focus for most prior research on group work in mathematics (with research on 'Complex Instruction' a notable exception, see, for instance, Boaler 2008; Cohen and Lotan 1997), a careful consideration of these findings can still inform our analysis of equity in schools. In the classroom context, I define equity as the *fair distribution of opportunities to learn*, where situated and sociocultural theories of learning are brought to bear on the question of what constitutes an opportunity to learn.

The term 'fair' here refers to a qualitative understanding of justice (Secada 1989), rather than a strict equality (i.e. in terms of counting the number of questions asked or the number of utterances for each student). What we consider to be fair is ultimately a political question as much as an empirical one; after reviewing theoretical and empirical research on learning in groups, I will return to the question of how I will define fair in the context of this chapter.

As we consider classroom opportunities to learn, it is important to specify first what learning is and the conditions under which it occurs (i.e. the conditions that create opportunities for learning). In mathematics classes, learning encompasses more than socializing students into particular mathematical practices; learning also includes changes to student identities. Thus, when analyzing opportunities to learn, we should consider not only students' *access to mathematical content and discourse practices*, but also their *access to (positional) identities* as knowers and doers of mathematics (Esmonde 2009). Of course, participation in mathematical discourse practices may influence one's identity in the classroom and one's identity may reciprocally influence how one participates in the group. These two aspects of opportunities to learn, and the relationships between them, are illustrated in Fig. 4.1 and will be discussed further below.

To elaborate on these two branches of opportunities to learn, and the fair distribution of opportunities to learn in cooperative learning, I will highlight instances as they arise in the following vignette.

> Tony, Sarah, Mustafa and Kendra are working together on a mathematics problem. From the teacher's vantage point at the front of the room, they look extraordinarily productive; their heads are bent together, they are engaged in animated discussion and Sarah's hands

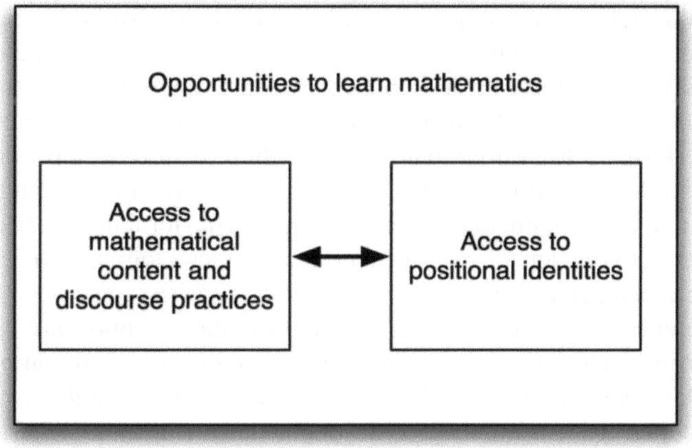

Fig. 4.1 Opportunities to learn includes access to (**a**) mathematical content and discourse practices and (**b**) positional identities

> gesture towards her own paper as well as Kendra's. As the teacher circulates around the room, she pauses to observe and listen closely to the talk. Tony and Mustafa are both writing, heads down, in their own notebooks. She hears Sarah explain her strategy for solving the problem, as she gestures toward her notebook and the diagrams she has written there. She hears Kendra tell Sarah, "Oh… I get it. But I did it a different way. What do you think? I chose …" and then notices Sarah's gaze drift back to her own paper as she begins working on the next question.

In this vignette, opportunities to make sense of mathematical ideas and to participate in mathematical discourse practices are not evenly distributed across the group. In cooperative groups, mathematics learning has been shown to be associated with specific types of interactions, including asking questions (King 1991), discussing problem-solving strategies (Chizhik 2001), observing someone else's problem-solving strategies (Azmitia 1988), explaining one's thinking (Nattiv 1994; Webb 1991) and maintaining joint attention (Barron 2000, 2003). In this example, Sarah does the explaining and Kendra listens, while Kendra does not have a chance to explain, nor Sarah to listen to an alternative strategy. The group as a whole does not maintain joint attention, as evidenced by Tony and Mustafa's independent work and Sarah's ignoring of Kendra's attempted explanation.

This is not to say that students in the group are not learning or will not learn from this interaction. Mustafa and Tony may be engaged in important sense-making processes and individual work may, at times, be a productive part of cooperative learning. An emphasis on participation in discourse practices should not be confused with an emphasis on talk alone. The point here is that for a group's interactions to be equitable, they should explicitly attend to the meaning-making processes of each of the group members, not just the few who are considered 'smart' or most competent.

Access to positioning as competent mathematicians is also distributed unevenly in this vignette of group interaction. Sarah is positioned as competent through Kendra's uptake of her idea, while Kendra's ideas are ignored. I understand identity through the ways that students position themselves and the way they position others (Davies and Harré 1990), rather than understanding identity as a personal or individual trait. Students are positioned in multiple ways in interaction, including positioning with respect to mathematical competence, behavioral norms, friendship and socially constructed norms of race, gender and socioeconomic status, as well as a host of other social categories.

When I refer to 'access to identities', I refer to students' opportunities to develop positive positional identities that place them as authoritative and competent members of the classroom community. This access occurs both through the moment-by-moment practice of positioning and longer-term trajectories of positioning (Wortham 2004), though the relationship of one moment of interaction to a person's longer-term trajectory of identity development is not always a simple one (Nasir and Saxe 2003).

In mathematics education research, issues of positioning, identity and identification are gaining currency. One particularly relevant study focuses on what Boaler calls 'relational equity' – when students of varying cultural backgrounds, gender and prior achievement treat one another's classroom contributions with respect (Boaler 2006a). Based on the pedagogical approach known as Complex Instruction (Cohen and Lotan 1997), the teachers in Boaler's study used complex, 'group-worthy' problems, emphasized the wide variety of skills necessary for success in the class and publicly recognized the important contributions of low-status students in the classroom. Over time, students began to treat one another with respect and to consider everyone's contributions to be worthy of serious consideration. In effect, the students in Boaler's study demonstrated how classroom positioning can shift over time, so that more students can be positioned as competent. Relational equity is encompassed in the broader construct of opportunities to learn as one aspect of access of positional identities as knowers and doers of mathematics.

1.1 Analysing Equity in Terms of Opportunities to Learn

The foregoing research allows me to elaborate more on the meaning of 'fair' as it relates to opportunities to learn in mathematics classrooms. To analyse group members' access to mathematical content and discourse practices, it is important to look at what I will call the group's *work practices*: interactional patterns that groups construct as they get work done together. In other words, work practices are a subset of classroom and group discourse practices – they are the discourse practices that the group invents, appropriates and directs towards a specific goal: accomplishing a task that has been assigned for their small group. These work practices may vary depending on the type of task and may also be quite different from the discourse practices they use in other types of classroom activities, such as whole-class discussions. Although the work of negotiation may at times be invisible, work practices are always negotiated among group members, in the sense that no one student can determine the group's patterns of interaction without the co-operation of others.

In my analysis of the presentation preparation, I will try to characterize the different work practices that the groups constructed together and then discuss the implications for equity. The key analytic question will be: *Based on the group's work practices, did all group members have access to mathematical content and participation in mathematical discourse practices?*

To analyse access to mathematical identities, I will investigate whether group members' positions allow them access to the mathematical ideas at play in the group (thus promoting mathematical learning), as well as whether group members' positions allow them to be perceived (at the moment or in the future) as mathematically competent and confident. The key analytical question is: *How do various acts of positioning influence access to mathematical content, mathematical practices and mathematical identities?*

2 Methods

The research was conducted in three different high school mathematics classes, all taught by the same mathematics teacher, Ms. Delack. (I will sometimes refer to the classes as Period 1, Period 4, and Period 6 respectively.) The three classes were all housed on the campus of a large, diverse urban high school, which I call Bay Area High School (BAHS). They were diverse with respect to race, gender, prior achievement and grade level.

Ms. Delack used the Interactive Mathematics Program (IMP) curriculum in all of her classes. This curriculum emphasizes conceptual understanding and open-ended problems and is explicitly designed for use with cooperative groups. With IMP, a day's work typically consists of one to three deep, rich problems, often with multiple solution paths or several 'correct' answers. The three classes that participated in this study were taking the Year 2 course, known as IMP2. Ms. Delack followed the IMP2 curriculum during most of the year, occasionally supplementing it with her own activities and materials.

2.1 Major Data Sources

The study was designed so as to obtain rich ethnographic data of cooperative group interactions and mathematics learning over an academic year, with a secondary goal of observing cooperative learning in more than one classroom.

Video recordings and ethnographic methods were essential to capture the details of talk, gesture and body positioning in interaction. Approximately 150 h of video were collected in all, though only a small sample of this data is reported here. Supplementary data in the form of student work, classroom artifacts, interviews and questionnaires were collected to support interpretation of the video data.

In Ms. Delack's classroom, students worked in groups on almost a daily basis. In these classes, 'group work' did not mean just one type of activity. In a single 55-min period, there were typically 3–4 different activities, including group discussion, presentations, whole-class discussions, quizzes, 'classwork' and 'homework'. For analysis I chose one activity structure – presentation preparation – that had seemed, in my visits to the classroom, to support a broad range of participation from students. In the three classes, students presented problem solutions in their class at least once a week. Ms. Delack required all students in the group to be prepared to present. Sometimes the group would choose the presenter and sometimes Ms. Delack would randomly choose someone just prior to the presentation.

Although group members seemed to participate more intently during this type of activity than during other, less-structured activities (e.g. a generic 'classwork' activity), without close analysis I could not determine the nature of these interactions: whether students focused on answers, procedures or conceptual explanations whether all students had opportunities to explain and ask questions; whether the interactions were dominated by one or more students. For this chapter, I analyse six examples of presentation preparation from the three classes.

As a way to identify differing patterns of interaction in the group, each video was viewed repeatedly and coding documents were created that answered the following questions about the group work practices.

- *How are ideas solicited?* Some groups read the materials together and students offered input, whereas in other groups students worked individually and responded to direct questions when asked. I noted the formal and informal participation structures within the group – roughly, who was included or excluded in activity and how (Goffman 1981).
- *How is correctness determined?* Once mathematical ideas were 'out on the floor', groups had to have some way to decide if those ideas were correct or not. In some cases, mathematical argumentation was used, while in others correctness seemed to be largely attributed to individuals (as in, 'smart' students were assumed to be correct).
- *Are there multiple correct answers/strategies?* This question was relatively straightforward to code, although brief descriptions were usually included in addition to a yes/no answer.
- *How are students positioned?* Although some aspects of positioning could be inferred from previous questions in this list, here I explicitly recorded how students were positioned with respect to classroom authority, including but not limited to mathematical competence.

2.2 Identifying Work Practices

Based on repeated viewings of the tapes, and iterative cycles of coding based on the questions listed above, several themes emerged in the groups' work practices and positioning. I was able to group work practices loosely into three basic types that were present in this data corpus: each segment of interaction could be coded as exemplifying an 'individualistic', 'collaborative' or 'helping' work practice. I will describe them briefly here to help the reader get a broad overview of the coding system. I will provide many illustrative examples of these three types of work practices in the next section of this chapter.

Individualistic work practice. Many groups went through periods of individual work during the activities. The distinguishing characteristics for interactions that I coded as individualistic were a propensity for: working individually before consulting one another; not asking for help when needed; denying help to group members who expressed confusion or requested assistance. Groups were *not* coded for an individualistic work practice if, for example, group members tended to work individually to carry out strategies that had previously been discussed in the group.

Evidence for an individualistic work practice would include: (a) seeing a group member ask for help and be refused (especially if the group had previously helped someone else or had displayed somehow that they were capable of helping); (b) a group member visibly struggling with the work without asking peers for assistance; (c) groups working alone with no discussion.

Collaborative work practice. I coded group work practices as collaborative when group members put their ideas together, worked together and seemed to act as 'critical friends' when considering one another's ideas – rather than quickly accepting or rejecting one another's ideas, collaborative groups discussed and critiqued ideas put out onto the public floor.

In collaborative interactions in which groups displayed confidence about a correct answer, several students would contribute to the correct answer or several different formulations or strategies were considered to be correct. Interactions in which groups faced greater levels of uncertainty or disagreement were considered collaborative if group members put more than one idea forward for discussion or if several people jointly constructed a single strategy.

Helping work practice. The 'helping' work practice was used to distinguish groups in which mathematical talk was asymmetrically organized, in which one or more students instructed other students about what to do. In contrast with the collaborative work practice, a helping work practice was characterized by the uncritical uptake of ideas and an insistence on just one correct idea or solution, while ignoring other possibilities (usually denying contributions from other group members once a correct answer had been found). Although it might seem counter-intuitive, the helping work practice could still occur in groups in which there were no expert students to provide the help. Groups were coded as 'helping' if group members oriented their actions towards obtaining an answer from a perceived more-expert other or if they gave up when there was no such expert available.

2.3 Identifying Positioning

As I will show throughout the analyses, differences in positioning for individuals in the group made a difference in opportunities to learn for all group members. For positioning, the major differences appeared to be associated with levels of authority in the group (Engle and Conant 2002). There were two types of authority that clearly influenced the nature of group interactions and affected opportunities to learn: whether or not a group positioned at least one group member as expert and one as novice, as well as whether or not a group positioned at least one member as having the authority to organize participation (e.g. to get the group started, to encourage particular kinds of participation, etc.).

Experts, novices and in-betweens. In many of the data examples analysed here, students positioned themselves with respect to mathematical competence. Students could position themselves as more or less competent than a peer (e.g. "I know less than you") or more or less competent at a particular task (e.g. "this is easy").

I used the code 'expert' for a group member who was frequently deferred to (mathematically) and who was often granted authority to decide whether their own and other students' work was correct. In order to be coded as an expert a student must have positioned her- or himself as such and must also have been positioned as an expert by peers. For a group to position one member as an expert, there must also be a student positioned as novice – with no novice, there is no one to defer to the expert and to take up their ideas.

When coding a particular interaction, I use the term 'novice' to refer to a student who deferred to an expert (positioning themselves as less competent) and whose opinion was frequently passed over in discussions of mathematical controversies (positioned by others as less competent). Novices were often instructed by others, and often accepted these instructions, though they sometimes challenged or questioned the expert's advice. However, disagreements between expert and novice were resolved quickly, usually by means of a simple assertion by the expert.

In some interactions, group members could be positioned as neither expert nor novice,[1] and students who were coded as expert in one situation may not be coded as such in another. In some interactions, there was one student positioned as expert, in others there was more than one and some interactions involved no experts.

Facilitators. The second type of positioning that will be considered here is that of the facilitator. I use the term to describe students who orchestrated the group activity and fostered broad participation from group members. In an interaction, a student positioned as facilitator made sure that all or most group members participated in group discussion in some way. This could take the form of making sure that group members asked questions when they needed help, of assigning different tasks to different group members or (in some of the most productive cases) of actively encouraging group members to contribute to joint problem solving.

[1] I sometimes refer to these students as 'in-betweens', in reference to Eckert's (1989) description of Jocks, Burnouts and In-Betweens in her high school ethnography.

3 Equity and Interaction in the Presentation Preparation

Figure 4.2 displays the negotiated work practices and positioning for the six examples of presentation preparations. This activity structure seems to allow for a variety of work practices as well as a range of classroom positions.

For the presentation preparation analysis, I begin with a general summary of how group interactions influenced equity and opportunities to learn, and then illustrate with examples of interaction segments that I allocated to the various cells of Fig. 4.2.

The strength of the presentation preparation is that its structure encouraged more than one person to get involved. For each presentation, multiple students in the group had to be involved either in preparing a transparency or poster to present or in rehearsing an explanation. Further, since the group typically did not know who would present, or knew that several presenters would be needed, it was possible for a number of students to rehearse explanations and to benefit from participating in this practice.

The weakness of this activity was that group members were able to opt out of discussing their ideas, listening to peers or participating in writing solutions. Although all group members were typically offered the chance to participate, many declined. In some cases, group members even declined to help their peers

PRESENTATION PREPARATION — WORK PRACTICES / POSITIONING	Expert and facilitator	Expert, no facilitator	Facilitator, no expert	No facilitator, no expert
Asymmetric: Helping		2 presentations 'Expert tells'	1 presentation Need outside help, facilitator organizes	2 presentations Stuck - ask teacher for help
Symmetric: Collaborating	2 presentations Everyone contributes			1 presentation Peers collaborate
Individualistic	1 presentation Little talk, one person writes	3 presentations Little talk, expert writes		2 presentations Stuck-individual asks teacher to explain

Fig. 4.2 Positioning and negotiated work practices for six presentation preparations

when asked. It appeared that students did not always hold themselves accountable to one another.

In this activity, the most equitable group collaborated to come up with a number of ways to present a solution (and it was coded as collaborative, expert present, facilitator present). One student acted as a facilitator (and also expert) and was able to encourage most students to rehearse presentations. However, this could have been improved if the group had focused more on connections among the varied ideas, rather than considering them one by one. The mathematical discussion also seemed to pass through the expert student, since group members spoke to the expert rather than speaking to one another.

The individualistic groups were much less equitable, in that students did not get the help they needed: they turned to the teacher for help rather than to one another and students opted out of participation. This may be because the presentation preparation was comparatively low-stakes for the students. Though the group knew that one or more of their group members would have to present, they were not graded on it and there was less interdependence built into this activity than it might seem. There were two reasons for diminished interdependence. First of all, presentations in these three classes often involved a student standing at the front and simply reading aloud what had been written on a prepared transparency or poster. This did not require a deep understanding of the material. Secondly, when questions were posed to the presenter, sometimes other group members would step in to explain. Thus, the presenter did not have to master the material she or he was presenting. The structure of the activity, as well as classroom work practices for presentations, allowed for a more individualistic approach to preparing presentations. As long as the transparency or poster had been prepared, the group could consider their task done.

Another interesting pattern was that many groups constructed several different kinds of work practices for different phases of activity. It appeared that for many groups, the work of preparing the transparency comprised one task and the work of rehearsing presentations was another. Some groups did not rehearse: in these groups, opportunities to learn were skewed heavily towards the person who prepared the transparency and any students who helped or collaborated with that task.

I now turn to examples of group interaction to add detail and complexity to the general characterization given above. Because there was such a wide variety of interactional styles represented in Fig. 4.2, I will not give examples of each. Instead, I focus on several examples that highlight the diversity within this particular activity structure.

1. *Individualistic work practice, with expert, no facilitator*
   In the first example, a student positions himself as facilitator but fails because his peers do not respond to his overtures; he is, however, acknowledged as the group's expert.[2]

[2] This example is presented in prose form rather than through a transcript, because large portions of the interaction occurred in silence.

Riley pulled the transparency closer to him and offered to carry out Ms. Delack's instructions. While he wrote on the transparency, Shayenne and Dawn chatted with one another about other matters, and Ayodele sat in silence. At times, Riley quietly verbalized what he was writing out loud. When he did so, he did not look at the other students and they did not visibly react to him. His tone indicated that he was talking mainly to himself. When he had finished writing a section of the transparency, he turned to Ayodele and asked him if he wanted to write a section. Ayodele said no, giving his bad handwriting as a reason.

In this example, the group allowed one student to do the work without consultation from the rest of the group. It was coded as individualistic, because Riley did the work without checking with other group members and because when he invited Ayodele to participate, Ayodele refused to do so. The mathematical ideas on the public floor were constructed by Riley, with no contributions from other group members. Other group members did not participate in this construction, even through listening to or reading Riley's explanations.

I consider Riley's attempt to engage other students in preparing the transparency as a bid to facilitate the group; since Ayodele refused (as the other two group members did at times during this group's joint work), Riley was not coded as facilitator. He was coded as expert, in this excerpt and throughout this group's interactions, because he was consistently positioned as knowing what the group should do mathematically and he was responsible for making sure that the answers on the transparency were correct. This interaction was inequitable, because the student who was positioned as most expert ended up doing all of the work, with no involvement from the other students, who were positioned as less competent and who may have needed to learn the material.

2. *Individualistic work practice, with expert, facilitator*
In this example, a group of four students had to prepare a transparency with answers to four different homework problems. A student positioned as facilitator suggested that they split up the work, so that each student was responsible for a particular piece of it (each of those students then being positioned as expert on their portion). This interaction was coded as individualistic because each group member wrote their own portion of the transparency, with no input from others. This interaction was equitable, in the sense that each student was positioned as expert and, therefore, as mathematically competent and able to represent the group. However, the group interaction could have taken greater advantage of their mathematical competence by discussing the problems to ensure that each student had opportunities to learn from one another.
3. *Individualistic work practice, with no expert, no facilitator*
In a later interaction, this same group was coded as individualistic, with no expert and no facilitator, because one of the group members became confused and asked for help on his problem. No group member agreed to help. In fact, they all positioned themselves as incompetent to help and did not display any sense of being accountable to help their group member. None of the students was positioned as expert for that small stretch of interaction. This interaction was inequitable in that a student who asked for help was denied any opportunity to engage in talk that might have supported the development of mathematical understanding.

4. *Helping interactions, with expert, no facilitator*
   Turning to examples of interaction that were coded as helping, there were two basic kinds: helping interactions in which one student was positioned as expert and interactions in which there was no expert. Helping interactions that included an expert tended to be quite similar to one another. In these interactions, the expert did most of the talking, with very little input from students positioned as less expert. Although it is, of course, possible for the non-expert students to learn from such interactions, the expert student had no way of knowing what non-experts were learning, because so little floor time was given to them.
5. *Helping interactions, with no expert, with or without facilitator*
   In helping interactions with no expert, groups had to find help from some other source. They invariably turned to the teacher, who was usually able to provide the information that the group needed to continue to make progress. The teacher's help was often provided in a more equitable fashion than the help of expert students, because the teacher usually asked questions of the students to find out what they had already done and to encourage them to make conjectures about how to solve the problem. If the teacher was not available, these groups tended to sit and wait for her.
6. *Collaborative work practice, with expert, facilitator*
   Finally, I present two contrasting examples of collaborative work practice. In the excerpt below, we see two students collaborating to construct an explanation for a problem that has already been solved. The problem involved trying to find the number of computations a computer could do in 30 s, if it could do one computation in $5*10^{-7}$ s.

1	18:29	Christa	Do you know how to explain this?$_1$ 1. (*points to the transparency*)
2	18:30	Tony	$_2$Do you get it? 2. (*turns head towards Candie, then back to the front, looking down at his calculator*)
3	18:31	Candie	(1*s*) $_3$Big numbers?$_4$ 3. (*reaches for transparency*) 4. (*takes the transparency that Christa hands to her*)
4	18:32	Christa	Yeah
5	18:33	Candie	Okay (*1s*) It's point zero $_5$zero zeozeozero fi::ve (*clicks tongue*) And then five times ten to the negative seven equals 5. (*Tony turns head towards the transparency in Candie's hand*)
6	18:40	Christa	(*points to the transparency*) $_6$This number right here$_7$ 6. (*points to a spot on the transparency*) 7. (*leans back in her seat*)

This interaction was coded as collaborative for two reasons: because Christa invited Candie to participate even after a correct solution had been accepted by the group and also because Christa and Candie jointly produced an explanation. (Even though the explanation was guided heavily by the transparency, Candie took the lead in producing it.) In this excerpt, in turn 1, Christa opened up space for collaboration by inviting Candie to participate. Tony echoed this invitation. Although this group had already prepared a transparency and could be considered finished with their task, Christa and Tony purposely invited Candie to co-construct an explanation to prepare for the upcoming presentation. In turn 5, Candie began an explanation, which Christa helped her with in turn 6.

Christa and Candie co-constructed the explanation, focusing mainly on explaining which arithmetical operations were used and how the different lines of the written solution corresponded to one another. Christa was clearly positioned as expert – she had written the transparency and she was charged with decoding it for others – and Candie, as novice. The explanation focused on the procedural level, rather than addressing what the problem was about and how to marshal mathematical reasoning to solve it. As this example shows, a collaborative work practice does not necessarily mean that students were engaged in particularly meaningful or intense mathematical discussion. The key difference between a discussion of this nature that was coded as collaborative, and one that was not, is that in collaborative discussions more than one student had the opportunity to voice developing understandings of the mathematics at play. Of particular interest, though, was the way that discussions tended to pass through and be dominated by an expert student – even if that expert were opening up opportunities for others.

In a contrasting example, one group interaction in the data corpus was particularly interesting, because students in it seemed spontaneously to construct various ways of displaying competence. They distinguished between *doing* a mathematical task, *understanding* a mathematical concept and *explaining* the concept. Students were able to position themselves as competent at some part of the task (usually, understanding) and less competent at another part of the task (usually, explaining). As in Complex Instruction, where the recognition of multiple abilities allows more students to be positioned as high status, the distinctions among doing, understanding and explaining allowed group members to be positioned as authorities while still expressing some uncertainty and confusion.

As an example of this distinctive positioning, in this group when one student (Candie) told her group, "I know what I'm doing a little bit? I just can't explain it, I guess," she was asked how she might explain it. She then attempted to give an explanation while the whole group listened and then was given feedback by the most expert student in the group. Rather than being marginalized and positioned as incompetent, she was offered an opportunity to voice her developing explanation. This contrasts with the earlier example in which, when Candie expressed difficulties with the presentation, her expert group member jumped in to help her complete her thought. In this second example, when Candie expressed difficulties with explaining, she was still positioned as competent at actually doing the problem.

These detailed examples have shown that there were a number of different ways that students could take up the presentation preparation and that the varying work practices and types of positioning influenced the opportunities to learn that were made available to group members.

4 Discussion

The goal of this chapter was to consider how the structure of an activity can influence a group's work practices and positioning. While an activity structure might encourage specific forms of engagement, all participants are constantly improvising within (and sometimes crossing the boundaries of) that structure, so some variation was expected (Erickson 1982). The examples given here illustrated a variety of ways that different groups could take up a single activity and demonstrated that these variations affected the group's distribution of opportunities to learn.

The most equitable groups tended to work collaboratively rather than individualistically or by helping. This is not simply an artifact of the coding scheme. Periods of individual work can support mathematical understanding, so long as group members can get help when they need it. And a helping work practice is not *a priori* inequitable. If a student who needed help was first asked to explain exactly what she or he was struggling with or to explain what was already understood, then the rest of the group could work from that basis to help. The problem with the helping practices in this particular study was that the helpers – positioned as unchallengeable experts – began with *their own understanding* rather than the understanding of the student who needed help.

This study therefore complicates research on the benefits of interaction with a 'more competent other' for learning. For example, many educators draw on Vygotsky's work to argue that working with more knowledgeable peers or adults can help students move into their 'zone of proximal development', thus enabling them to develop more sophisticated mathematical understanding (Kozulin et al. 2003; Wertsch 1984). For example, Wertsch (1985) provides a detailed analysis of adult–child interactions and demonstrates that adults are often quite sensitive to the children's levels of understanding, and they can subsequently work to develop a common language with which to communicate.

By contrast, in this study, helping interactions in which one student was positioned as more expert than the others were sometimes found to be detrimental to learning. The experts in this study did not, for the most part, act like the adults in Wertsch's study or like their own teacher. They did not actively try to understand and build from the novice's perspective, which was a technique that their teacher usually used when working with groups. Many of the expert–novice interactions provided a fairly narrow window on mathematical content, with the expert student focused mainly on conveying her or his own mathematical ideas without considering the ideas of the novice students. Some novice students actually set aside prior understanding and displayed less mathematical competence than they had prior to their interaction with the expert. (Further details about this context appeared in Esmonde 2006.)

The findings here resonate with much prior research in the field of cooperative learning, while extending and deepening the discussion of equity in cooperative classroom contexts. For example, this study echoes and adds complexity to the finding that activity structures with interdependence and individual accountability can sometimes increase student achievement. This study goes beyond prior work in this area because of the careful analysis of the relationship between the activity structure and student interactions.

A second finding from prior research is that those who seem to have greater knowledge prior to a group interaction often benefit the most from group work (Cohen et al. 1999; Fantuzzo et al. 1992; Webb 1991). This study suggests why that may be the case. In groups with experts and no facilitator, the expert students were likely to take on the bulk of the work, focus much of the interaction on explaining their own ideas while neglecting the understanding of their peers.

Of course, the most pressing questions for educators will be how to marshal the evidence presented here to improve equity in cooperative learning in their classrooms. We must exercise caution when applying the results of an ethnographic study, deeply rooted in a particular classroom community, to more broad contexts in education. However, this careful examination of a particular classroom context does provide clues as to how to structure cooperative learning more equitably.

As I noted above, the most equitable groups constructed work practices focused on collaborating, rather than helping or working individually. Groups without experts who worked collaboratively still made progress in mathematical understanding and groups with facilitators were able to take advantage of an expert's understanding, without compromising opportunities to learn for those positioned as less expert. So, finding ways to support more collaborative interactions among students should be a priority for mathematics educators.

Still, it would be a misinterpretation of the analysis presented here to suggest that there is one best model for group interaction or one type of positioning that all students should take up. There are at least two reasons why we should not assume that what I have characterized as 'collaboration' is always the best work practice. First of all, a number of studies have found that different styles of interaction may be appropriate to help students learn different kinds of material (Chizhik et al. 2003; Cohen 1994; Damon 1984). For more procedural types of learning, a helping interaction may be perfectly adequate – though other studies have found that helping interactions are more helpful when the recipient of help has the opportunity to state her or his own ideas, and to carry out independent problem solving after being helped (Webb and Mastergeorge 2003). And, of course, students might benefit from periods of individualistic learning, as they test out new ideas or try to consolidate ideas that they were introduced to in group collaboration.

Secondly, students' preferences for and interpretations of particular work practices may be related to their repertoires of practice, developed through participation in communities inside and outside of school (Gutiérrez and Rogoff 2003). A study in a Dutch middle school found that immigrant students were more likely to construct (what I would call) helping interactions, while Dutch-born students were more likely to construct (what I would call) collaborative interactions. Further, each group tended to consider the work practices of the other group to be disrespectful (de Haan and Elbers 2005). Thus, cultural background, communities of practice and probably age, subject matter and experience in the classroom all affect how students work together. It would be a mistake, then, to try to impose some *a priori* set of work practices and classroom positions on students or to think that only one way of learning together could be productive. Instead, we must find ways to capitalize on and benefit from the diversity of students' approaches to learning (Nasir et al. 2006).

Consider the presentation preparation. Groups that were the most equitable took advantage of this activity to allow students to practice giving explanations. Multiple correct explanations were sometimes encouraged, providing group members with a window into one another's thinking. However, some groups did not rehearse any explanations at all, satisfying themselves with only a written representation of a solution, prepared by a single group member. One could alter the structure of this activity by requiring that each student rehearse an explanation and each group give feedback to one another on these rehearsals.

Another possible way to foster more equitable cooperative interactions, discussed in depth in Boaler and her colleagues' recent papers (Boaler 2008; Boaler et al. 2006; Boaler and Staples 2008), is to change the way expertise gets assigned and defined in the classroom. Recall that in one group students were able to position themselves as experts in one domain (*understanding* how to find the feasible region) and as less expert in another (*explaining* how to find the feasible region). This assignment of competence appeared to open up possibilities for interaction that were not often seen in the data corpus. The crucial difference seemed to be that students who expressed uncertainty about explaining, but who said they understood, were asked to give a practice explanation. They were then supported in constructing an explanation. This contrasts with the work practices in many other groups in the data corpus, in which students who expressed uncertainty were often positioned as 'not understanding' (as opposed to 'understanding but not explaining well') and were often given more directed explanations – explanations which often did not shed much light on the mathematical phenomenon in question.

Of course, changes to the activities would require further research, as classroom interactions are complex and we cannot always predict how groups will react to the shifts in structure. As I have tried to show here, one can set up a co-operative activity, but what happens within that activity depends on the choices that are made by participants.

A second direction for future research would focus more explicitly on issues such as race, gender, (dis)ability, language and socioeconomic status. Although I do not do so in this chapter, it should be a priority to build on this approach to consider how race, gender, socioeconomic status and other social categories play out in terms of participation. The inequities that we see in the world outside of school are often replicated within schools and within cooperative groups. This study suggests several important questions about how this process works. *How do work practices get formulated and what is the basis for student positioning? Can it be shown that patterns of interaction within small group work systematically privilege or marginalize students from certain groups? Is positioning with respect to mathematical competence related to positioning of other kinds – with respect to race, gender and other socially constructed identities?* The approach presented in this chapter discourages essentialization of marginalized groups and instead offers a way to analyse these systemic inequities through participation patterns, by focusing on micro-interactions in classrooms that are consequential for opportunities to learn.

Note

The material in this chapter is based on my article 'Mathematics learning in groups: analyzing equity in two cooperative activity structures' (published in 2009 in the *Journal of the Learning Sciences* **18**(2), 247–284) and my doctoral dissertation completed at the University of California, Berkeley, under the direction of Dr. Geoffrey B. Saxe. This work was supported by the National Science Foundation under Grant No. ESI-0119732 to the Diversity in Mathematics Education Center for Learning and Teaching, and Grant No. SBE-0354453 to the Learning in Informal and Formal Environments Science of Learning Center, as well as a Graduate Student Fellowship from the Institute for Human Development at the University of California, Berkeley. Any opinions, findings, and conclusions or recommendations expressed in this material are those of the author and do not necessarily reflect the position, policy or endorsement of the National Science Foundation or the Institute for Human Development.

I would like to thank Geoffrey B. Saxe, Alan Schoenfeld, Patricia Baquedano-Lopez, Samuel R. Lucas, Victoria M. Hand, Joseph Flessa and anonymous reviewers for their careful reading and thoughtful critique of the dissertation and earlier versions of my article.

Acknowledgement I am grateful to Taylor & Francis for permission to reuse material from the above-mentioned article. Such material is reprinted by permission of the publisher (Taylor & Francis Ltd, http://www.tandf.co.uk/journals).

Chapter 5
Aiming for Equity in Ethnomathematics Research

David Wagner and Lisa Lunney Borden

> Walking along a forest path with a Mi'kmaw teacher, I listen as he tells me about mathematics done in his community. Trying to be helpful, he asks me yet again, "Is that what you want? What else do you want me to say?" In the background, I notice his wife using mathematics without fanfare to measure the depth of a puddle for their son who wants to jump into it.

The above narrative comes from Wagner's reflections on a conversation intended to inform mathematics teaching in a Mi'kmaw community on the east coast of Canada. It was the first of many conversations involving various people in this Aboriginal community, all of whom had some relation to mathematics learning and a stake in the cultural issues at play in the community. This initial conversation and the ones that followed it illustrate that there are multiple actors involved in any mathematics learning situation and the form of their interaction relates closely to equity concerns. It is never straightforward to understand how these actors relate to each other in the development of mathematical ideas. In this chapter, we will describe the development of our interactions, which were motivated by our concerns for Mi'kmaw students doing mathematics with little connection to their culture. In these interactions, we found ourselves increasingly attentive to discourse patterns and we intentionally shifted our positioning within the community in response to what we noticed.

Before considering three interrelated series of interactions that illustrate the development of our relationships in our research, we will describe the key scholarship

D. Wagner (✉)
Faculty of Education, University of New Brunswick, Fredericton, NB, Canada
e-mail: dwagner@unb.edu

L. Lunney Borden
Faculty of Education, St. Francis Xavier University, Antigonish, NS, Canada
e-mail: lborden@stfx.ca

B. Herbel-Eisenmann et al. (eds.), *Equity in Discourse for Mathematics Education: Theories, Practices, and Policies*, Mathematics Education Library 55,
DOI 10.1007/978-94-007-2813-4_5,

from which we draw, in addition to the nature of the Mi'kmaw people's marginalization, especially in relation to mathematics learning. The three sets of interactions relate to attempts to address this marginalization. The first set we discuss involves ethnomathematical conversations with elders and community leaders, aiming to uncover mathematics at work in the communities. The second arose out of our critique of the first situation. For this, we draw on examples from student ethnomathematical engagement and the instructions they received for doing this work. Thirdly, we reflect on connections between the ethnomathematical conversations in the community and others outside. As part of this reflection, we analyze excerpts of mathematics texts that demonstrate an overt desire for cultural sensitivity. These three accounts of interactions comprise our reflection on our roles as researchers bringing our agenda into the communities as part of our response to encouragement from the communities to work together to address mutual concerns for the children.

1 Context

The lack of interest in mathematics among Mi'kmaw youth has been a long-standing concern in Mi'kmaw communities. While it is difficult to gather accurate statistics on the number of Mi'kmaw students pursuing educational paths involving mathematics and the sciences, community leaders recognize and articulate concern about the disengagement of their students from these subjects. More generally, interested parties across Canada have expressed concern about the relatively low participation of Aboriginal students in mathematics- and science-based post-secondary programs. The Canadian government's national working group on education has said that a key area to be addressed in Aboriginal education in Canada is the development of culturally relevant curricula and resources in the areas of mathematics and science, where there is currently an identified weakness (INAC 2002). Although not specific to Canada, an NCTM publication also identified this need, saying that Aboriginal people in North America have the lowest participation rates of all cultural groups in advanced levels of mathematics (Secada et al. 2002).

Ezeife (2002), Secada et al. (2002) and others have identified a key reason for the disengagement of Aboriginal youth from mathematics and science – the discrepancy between their own cultures and the cultural values embedded in school-based mathematics programs. Cajete (1994) stated that when science is taught from a Western cultural perspective, it acts in opposition to the values of traditional culture for Aboriginal students, which affects their performance in mathematics and science because it simply is not connected to their daily lives. Lunney Borden (2010) has shown that the lack of attention to value differences and the use of inappropriate pedagogical strategies to be among the factors that result in a disconnect between school-based mathematics and Mi'kmaw ways of reasoning mathematically. As a result, many children choose to opt out of mathematics because the cost of participation is too high, demanding that they deny their own world-view in order to participate in the dominant view of mathematics. Doolittle (2006) and Gutiérrez (2007a), each in their own way, have elaborated on this cost of participation. The incidence of conflicting world-views has led many Aboriginal students either to ignore the possibility of studying science or mathematics or to struggle within these disciplines. This disengagement is a serious issue for Aboriginal communities that look to younger generations to acquire the skill and knowledge needed to move their communities closer to the realities of self-government in this modern age.

We note that disengagement goes both ways. As Canada's majority culture continues to marginalize Mi'kmaq[1] and other Aboriginal peoples, these marginalized peoples reject many of the dominant discourses of the majority. Individuals in Mi'kmaw communities could also be said to be ignoring, moving away from or marginalizing mathematics because of the cost of participation, just as the forms of mathematical instruction leave their needs unaddressed. When a dominant culture positions a community in a way that marginalizes the people, the people in that community, in response, may resist engagement with the dominant organizations and people. There are various ways of resisting, however, including spurning dominant cultural values or transforming aspects of the dominant culture's modes of promulgating its values and associated positionings. Some of the dominant culture's values are closely connected with mathematics education – for example, the privileging of mathematical knowledge and the kind of objectivity that is suggested in mathematics.

2 Ethnomathematics

Our research efforts have been aiming to address the disconnect between Canada's dominant culture and Mi'kmaw communities, as described above, particularly as this disconnect relates to mathematics education. An aspect of this work has been to engage in ethnomathematical conversations within Mi'kmaw communities.

Most important to us, ethnomathematics positions all mathematics as being culturally contingent. School mathematics responds to needs and problems that have arisen in particular cultures (usually not Aboriginal traditions, which are rooted in close connection to the environment), just as mathematical practices in Mi'kmaw communities respond to needs and problems in particular times and places with particular values.

Gerdes (1997), in his survey of the first decade of ethnomathematics, highlighted its way of uncovering mathematics in communities that are unaccustomed to recognizing the mathematics in their practices. Ethnomathematics can thus be seen to have

[1]Like Orr et al. (2002), we use 'Mi'kmaw' adjectivally and 'Mi'kmaq' nominally following the usage adopted by the Atlantic Canada Mi'kmaw/Miigmao Second Language Document (NSDOE 2002). Applying Mi'kmaw grammar within written English is not straightforward.

emancipatory power, because the uncovered mathematical practices can inspire confidence in students who may assume they cannot do mathematics. Likewise, we hoped and continue to hope that as Mi'kmaw children learn to recognize mathematics in their cultural practices they would be more likely to expect success in mathematics. Furthermore, we believe that ethnomathematics can make them better equipped to understand school mathematics by making connections between it and their cultural practices.

Since Ubiratan D'Ambrosio coined the word 'ethnomathematics' in the early 1980s (for his early writing on this topic, see D'Ambrosio 1985), it has become established in mathematics education research and has also been subject to significant criticism. D'Ambrosio (e.g. 1997) himself has raised criticisms, which relate mostly to the way ethnomathematics is received and, thus, by implication to the way ethnomathematics research is done and presented – for example, "Much of the research in Ethnomathematics today has been directed at uncovering small achievements and practices in non-Western cultures that resemble Western mathematics" (p. 15).

With a criticism similar to D'Ambrosio's, Dowling (1998) has challenged Gerdes' claim for emancipation. In describing the 'defrosting' of mathematics frozen in a woven button, Gerdes (1988) had celebrated the mathematics that was already present in Mozambique. He had claimed that ethnomathematics "stimulates a reflection on the impact of colonialism, on the historical and political dimensions of mathematics (education)" (p. 152). Dowling responded, calling this an example of the "myth of emancipation", noting that the "difficulty is that it appears that a European is needed to reveal to the African students the value inherent in their own culture" (p. 12) and that this revelation is to be done in European terms.

This critique weighed on our minds in the development of our conversations amongst the Mi'kmaq. Indeed, our initial conception of the research had the potential for the problems that D'Ambrosio and Dowling warn us about. In our account of the shifting storylines, we will answer Dowling's criticism of ethnomathematics as seemingly requiring a Western arbiter. The problem identified by D'Ambrosio, that small achievements are compared to Western mathematics, is not so easily addressed. However, we will address this criticism in our account of shifting storylines as well.

3 Positioning Theory

Positioning theory has provided us with a framework for critiquing our interactions in the research. Our sense of positioning theory follows the social psychology work of Harré and van Langenhove and its consideration by Wagner and Herbel-Eisenmann (2009) in the context of mathematics education. In Harré and van Langenhove's (1999) edited book, the general description of *positioning* refers to the way people use action and speech to arrange social structures. This positioning theory claims that, in any utterance, clues in the word choice or associated actions evoke images of known storylines and positions within those stories. For example, as researchers, anything we say or do evokes certain storylines and the people with whom we interact may comply with or resist a storyline we initiate by responding in expected or unexpected ways.

In their contribution to the Harré and van Langenhove book, Davies and Harré (1999) focused interpretive attention on 'immanent' practices, in contrast to the common scholarly focus on 'transcendent' discourse structures. Using Saussure's distinction between *la langue* ("language") and *la parole* ("speech"), they differentiated between the practice and the system of a discourse in which the practice is situated, claiming: "*La langue* is an intellectualizing myth – only *la parole* is psychologically and socially real" (p. 32). This approach helped us map out the many people connected with our actions as researchers. Temporarily forgetting about the discipline of mathematics, the cultural practices of Mi'kmaw people and our goal of bringing these forces together helped us focus on the actual people and the interactions.

Though this approach focuses attention appropriately on human interaction, we note with Wagner and Herbel-Eisenmann (2009) that myths are the stories people live by: they thus have power and are, in this sense, real. For example, in this chapter we consider as real the discourses of mathematics and of cultures in conflict in colonialism, though positioning theory may seem to encourage us to ignore their force. Following the argument of Davies and Harré, however, we recognize that human interactions are more real than discourses, in the sense that they are more local, alive and dynamic; they are relatively receptive to a participant's contributions through action and speech. This view highlights the possibility of alternative structures of interaction.

Taking seriously the existence and force of mathematics, Mi'kmaw and European (Western) culture and colonial history, though they are transcendent discourses, helps us to identify storylines at work in our research interactions. One chapter in the edited book on positioning addressed the production and use of stereotypes, but it is, even by the authors' admission, not very developed. In that chapter, van Langenhove and Harré (1999) explained that social psychology (the field in which the book theorizes positioning) does not address stereotypes well. They recognized that stereotypes appear to be positions or characters in storylines, and that these stereotypes might be changed on a local basis by taking up new storylines, but they admit that they have no recommendations about how this might be done on a large scale. We see our efforts to shift the nature of our positioning in our research interactions as an example of the development of new storylines.

Any discourse is static in comparison to the dynamic possibility available to individuals and collectives in any instance associated with that discourse. Thus, the only available site for transforming a discourse is in individual interactions in the moments of action. The discourse is constituted by the sum of its many interactions. And so, we claim, there is emancipatory power in focusing on the real interactions of any moment and ignoring transcendent discursive systems. The following accounts of our interactions in and relating to the Mi'kmaw communities considers the nature and challenges of this emancipatory power.

4 Shifting Storylines in the Research Conversations

The opening account in this chapter comes from the beginning of our ethnomathematical field work. We had invited a particular Mi'kmaw leader and his family to walk with us in the forest to talk about mathematics practices (both traditional and current) in their community. He was trying to be helpful by telling us what we wanted to hear. We were grateful for this spirit of cooperation because, according to our planning, it would help us create culturally-appropriate resources for students in his and other Mi'kmaw communities. However, we were a little disturbed that he kept asking if he was saying what we wanted to hear. On reflection we recognized two concerns. Firstly, we did not see ourselves as the ultimate audience of his observations, yet he and we together had positioned us as his audience. Secondly, we worried about authenticity, because he seemed to be subordinating himself to our agenda and we did not talk about his agenda(s) at this time.

Further, it was interesting that he was talking about mathematics in his community, while his wife was in the background doing mathematics. We were listening to talk about mathematics and apparently ignoring mathematics in action. The leader's wife had used a stick to measure the depth of the water and compare that depth with the height of her son's boots to demonstrate for him the foolishness of his wish to jump in the puddle. She had said nothing during this episode and very little in our long walk together. Nevertheless, her non-verbal message had been heeded by the boy.

Thinking about our conversation in terms of participants (using the lens of positioning theory), we envisioned something like the diagram in Fig. 5.1. In it, we refer to the teacher from the situation described above as a community representative. His status as a representative of the community came from at least two distinctions. He held community honours that recognized his knowledge of traditions. He was also respected as a teacher who understood the traditions and values of the dominant culture and who was thus well-equipped for intercultural interaction. The people we refer to as being outsiders include a wide range of people, including scholars who would read our research reporting, teachers in Aboriginal schools for whom we would write accounts of the mathematics we would illuminate and Aboriginal students who would be exposed to these accounts through their teachers who will have read of them and through materials generated by the research.

We also had ethnomathematical conversations with other community leaders, including elders. The diagram reflected the interaction patterns for any of these conversations.

In the diagram, we highlight (in gray) our position as researchers to indicate our privileged authority. The teacher was telling us what *we* wanted to know and

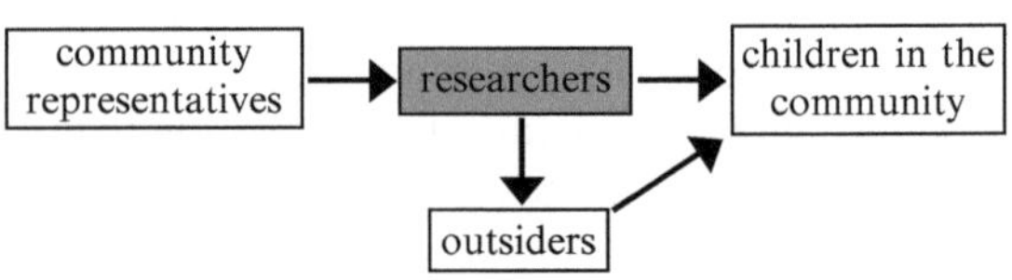

Fig. 5.1 Initial ethnomathematical interaction

reminded us regularly of this fact with explicit questions, but most often with his eyes and his expectant pauses interspersed through the sharing of his knowledge about traditional practices that could be deemed mathematical. According to the storyline that we were mutually constructing, we as researchers would decide what and how to pass the knowledge on to people outside the community and to the children within the community.

It is important to note here that this teacher and others in the community welcomed this research, trusting our judgment about how the community could best accomplish its general wish to make mathematics more relevant to the children of the community. This level of trust is not easy to come by in the Aboriginal communities, which have suffered much even from well-intentioned research and well-intentioned colonisation.

For example, the government policy White Paper entitled *Statement of the Government of Canada on Indian Policy* in 1969 purported to be acting in support of Aboriginal people, but ended up creating harm. This document claimed to have consulted Aboriginal people in an effort to create policy that would allow for "full, free, and non-discriminatory participation" (DIAND 1969, p. 5). Yet this policy was perceived by Aboriginal people as an effort to eliminate treaty rights. It prompted a response commonly known as the Red Paper that claimed they felt "stung and hurt by [the Minister's] concept of consultation" (Indian Chiefs of Alberta 1970, p. 2) and argued that the recommendations of the white paper would harm Aboriginal people. The red paper response demonstrates vigilance within the communities with respect to interventions from outside the communities and claims of consultation.

Another policy that claimed to be helping Aboriginal people was that of residential schools, yet these schools caused considerably more harm than good and negatively impacted the larger Mi'kmaw community (Knockwood 1992). As Battiste (2000) has stated, "these schools broke relationships among the people with themselves, with their own guardian spirits, their parents and communities, as well as with the land and environment" (p. 4). The trail of government decisions relating to policy regarding residential schools is outlined in Milloy's (1999) book *A National Crime.* These experiences and others are behind the communities' requirement that research within the communities be reviewed and formally approved by a council of Mi'kmaw leaders. We are honoured to have had our research approved by this process and informally approved by the ongoing relationships that have been central to the research.

Though we had approval for the kind of research with which we began, it was not our intention to be controlling. Though in any situation every participant has the opportunity to exercise agency, the way we positioned ourselves at the centre of the conversations described here positioned other people in roles that seemed to have limited choice – primarily the choice to follow our storyline or not, complicity or resistance. The storyline we initiated follows the Gerdes (1988) rationale, described above, and was approved by key people in this Mi'kmaw community. Though this situation generated some interesting revelations (see, for example, Wagner and Lunney Borden 2006, in press), the enacted storyline, to our embarrassment, was reminiscent of our region's colonialist history, which is a distasteful one.

Yet again, outsiders and their agenda are welcomed amongst the people of a generous and patient community, taking what they want from the people.

As with any situation, this one was complex because we were not necessarily seen as outsiders. Lunney Borden, who had worked in this community for over a decade and who was learning the local language, Mi'kmaq, was taken as an insider more often than she was taken as an outsider. However, this conversation was Wagner's first in the community. Lunney Borden bringing him in complicated her position as an insider. Together we were positioned as representatives of an institution (the university and academia in general), while at the same time being taken for who we were as individuals, a well-known ally and her colleague, one whom she trusted.

As predicted by Harré and van Langenhove (1999), attending to positioning opened up new opportunities. Our critique came to a point of action when the two of us were talking about our undergraduate teaching and noting that too often our assignments have us doing most of the thinking for the students: we preferred assignments that would have students doing the conceptual work as much as possible. The parallels between our work with our undergraduate students and our research work became obvious, and thus suggested to us that we were positioning the community's students as our students, for whom we were accepting some responsibility. Further, why should we do all the ethnomathematical work? The concern was not to limit our work, but rather to give others the opportunity to benefit from doing conceptual work that we had been doing following a model of ethnomathematics and to position community insiders as most responsible to each other with children and others in the community responding to each other.

Reflecting on Morgan's (1998) research that underscores the importance of audience in students' mathematical writing, we realized that positioning the children as the ultimate audience in a chain of knowledge sharing affords them no opportunities to address an audience other than their teacher and certainly provides no imperative to engage in real problems and issues faced by their community. It became clear that we should remove ourselves as medium of the transfer from elder to children. New storylines were necessary.

4.1 Changing Storylines

From this critique, we connected to a relatively new storyline in Canadian Aboriginal communities. As we agreed about the necessity of positioning Mi'kmaw children as collectors as well as receivers of knowledge, Lunney Borden identified a potential medium for the children's knowledge exchange. As part of the long-standing tradition of storytelling in Aboriginal communities, elders and others have recently begun to share stories and other forms of knowledge among communities across the country in 'contests', using the internet and real-time video conferencing.

Drawing on this storyline, we invited teachers and elders from some of the Mi'kmaw communities to gather and plan such a contest for promoting and exchanging students' ethnomathematical work. The "Show Me Your Math" (SMYM) contest, which was developed in this conversation, has now prompted over a thousand students from four provinces over a 3-year period to undertake ethnomathematical investigations to show others the mathematics used in their communities.

In order to break the school tradition of students doing work for teachers as audience, the teachers and elders who came together agreed to develop a video prompt comprising Aboriginal people inviting students' participation and describing the parameters of the contest. It featured only Aboriginal people, including an elder, a middle-aged teacher and children, all asking the viewer (the student) to "show their math." The elder featured in the video was party to some of our initial ethnomathematical conversations and also part of the group who gathered to develop and plan the SMYM contest. (The video prompt is available at: http://showmeyourmath.ca.) It begins with the elder, sitting in a classroom talking about mathematics. He says:

> What is Mathematics? Some people say it's what we do in math class or maybe what mathematicians do; but mathematics is much more than this. A mathematician named Alan Bishop has said that mathematics is counting, measuring and locating. When you design, explain or play with counting, measuring or locating, you are doing math. If you think of mathematics in this way, you might begin to see it all around you.

This introduction is followed by community representatives noting possibilities for projects. These include an 8-year-old boy saying, "I'd like to ask my Grammy how to say 'an oval' in Mi'kmaq"; a women saying, "I'm a plumber, I use math all the time"; a 14-year old playing music on his guitar and saying, "Math is in music. I would like to find out more about that"; a middle-aged male teacher saying, "I would like to see some students [looking at how government is] making decisions using math as a tool". The video ends with the elder who introduced it saying, "Now, show me your math", followed by two elementary-aged children repeating with gusto, "Show me your math!"

In response to this prompt, school children interviewed elders, experts in crafts and others to explore mathematics that has been used in their communities' traditional practices and also more current mathematics in their communities. In some schools, elders and other experts were invited by teachers into classrooms. In other schools, students interacted with community members outside of school. Students published their work on the internet site used for the other 'contests' on which the SMYM contest was modelled. Students also presented their work to the region's communities in a math fair. (For more detailed descriptions of student projects, see Wagner and Lunney Borden 2011.)

Figure 5.2 represents our view of this set of conversations, again using the lens of positioning theory to focus on the interactions among individuals instead of on the powerful cultures at play, including mathematics and Mi'kmaw traditions. Because the web of interactions in this set of conversations was much more complex than our initial ethnomathematical conversations, it was harder to represent in a diagram. We had much less control and access to the relevant conversations, and there were significantly more conversations that related to the web.

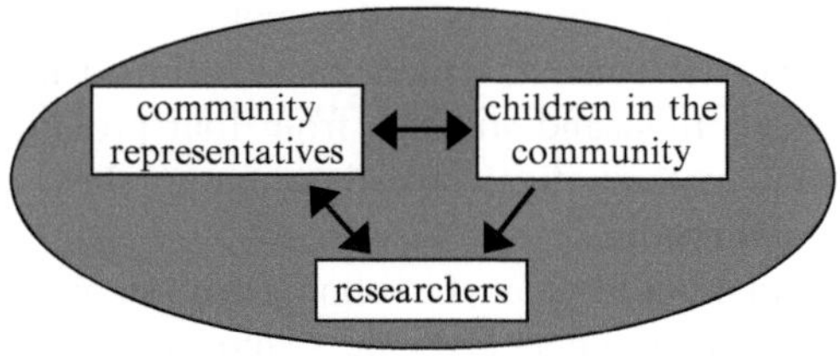

Fig. 5.2 "Show me your math" interaction

As researchers, we positioned ourselves in reciprocal relationships with people in the community by setting the conversation in motion. Our aim was to remove ourselves as much as possible from the many conversations and try to observe as much as possible. In this cloud of agency (all actors are in a gray cloud in the diagram), there were multiple conversations, each of which included the negotiation of intentions. Elders and other representatives of the communities had things to tell their communities' children. Children wanted to listen and it became obvious that the more they heard, the more they wanted to hear. We, as researchers, wanted to hear what elders, children and others valued in their conversations and we were interested in the collection of ethnomathematical explorations being compiled by students. The children and others in the community eagerly accepted the invitation for them to talk to each other.

Critiquing the web of interactions, we found ourselves once again most critical of our interactions with community representatives. We wrote the script for the elder to introduce the video. Thus, in a way, the video bears our words with his voice and face. However, the elder was not a mere front. He had been part of some of the initial ethnomathematical conversations and had demonstrated his acceptance and understanding of our account of ethnomathematics by giving multiple examples of mathematics at work in traditional practices. He was also part of the group who met to develop the SMYM contest and had recommendations for the development of the video. This group asked us to make the video.

While this elder had been far from passive throughout these conversations, he was also complicit, suggesting his approval of our actions. We note that such complicity is a form of agency. He did not have to disagree to express agency. In fact, different cultures express active support in different ways; his form of support was expressed in a culturally appropriate way. He made more concrete suggestions and provided more relevant information than anyone in the group developing the contest. For example, one of the most exciting aspects of this planning came when he explained for us all some of the different ways of describing a circle in Mi'kmaq. There are many Mi'kmaw words for circle, but none of them translate directly to the noun used in English – 'circle'. Rather, as is often the case in this and other Indigenous languages, there are verbs that relate to the idea that is represented in English by a noun. In this case, some of the ideas associated with a circle translate roughly as "it goes around" or "it is turning around". By contrast, in English it is natural to think of a circle as static and abstract because 'circle' is a noun. This part of the discussion prompted the inclusion in the video of the boy saying he wanted to ask his grandmother for the Mi'kmaw word for 'oval'.

Though we brought ideas to the community, we were acting under the direction and invitation of the community, including this elder. Our actions included writing a script for the introduction and omitting in the video some of the fascinating contributions from this elder and others in the planning. In the planning, we all agreed, after considerable discussion, that we would give brief examples to invite children to get their elders talking instead of deciding as experts what the students should be hearing. Thus, the decision to omit elaborated examples in the video was a group decision.

The examples of ethnomathematics provided in the video were roughly outlined partly by our choices of who to ask to appear in the video and more so by us telling these people why they were asked to suggest examples. Unlike the elder's opening statement, their words in the video were not scripted. Each statement related to the participant's own life experience or interest.

The script we wrote for the elder is also interesting in terms of the critiques of ethnomathematics we outlined above. The script includes what Prince et al. (1982) called an *attribution hedge*, which is any way of using language to shield oneself from critique by attributing a proposition to someone else. Rowland (1995) considered this and Prince et al.'s other types of hedges in his analysis of mathematics dialogue. The elder's definition of mathematics borrows authority by attributing the idea to Alan Bishop, someone unknown to most of his listeners yet with apparently strong credentials: he is described as a mathematician from the other side of the world and he has a white-sounding name, one which invokes church imagery (bishop).

Considering this attribution hedge, we address Dowling's critique of ethnomathematics, which we outlined earlier in this chapter. Do Mi'kmaq need a white mathematician to tell them that their community practices include mathematical activity? We think the answer to this question is yes, because mathematics itself as a construct is external to the community. Though there is much evidence of mathematical reasoning and problem solving within the historical and modern practices of Mi'kmaw communities, we note that there is no Mi'kmaw word for 'mathematics'. There are words to describe mathematical processes such as counting, measuring, navigating and designing to name a few, but these were not initially seen as mathematics by the community members we spoke with in our earlier conversations. These cultural practices were evaluated according to how they address community needs, not in terms of mathematical values.

Mathematics is something brought into the community by outsiders through colonialist education (from past to present schooling). Thus, because mathematics is seen as being held (owned and represented) by outsiders, having an outsider with credentials release this hold opens up this field of study to invite Mi'kmaw people to contribute their ideas, approaches and connections to the field of mathematics. This also relates to D'Ambrosio's criticism of the way ethnomathematics often focuses on small achievements in a culture and implicitly evaluates these achievements in terms of their connections to Western mathematics. There is no need for the word 'mathematics' in Mi'kmaw culture, except for its presence in the school system. Thus, identifying community practices as mathematical by implication connects them to what happens in mathematics classrooms. In this way, local cultural achievements are related to Western mathematics practices.

Along with the release of a colonialist hold on mathematics, we see greater significance in the students' need to be invited into a new way of seeing mathematics by a local elder, who releases them in another way. We described above how marginalization goes both ways. The elder in the video invites his community's children to connect the community's practices, which are very dear (and some even sacred), to mathematics, which has been connected to colonialism. The invitation structures a relationship in which the students address a local audience, contrasting the usual classroom interactive structure that positions the teacher as the audience. The teacher, in such a relational structure, represents another culture's knowledge and values, that of the prescribed curriculum and measures of achievement, even if the teacher is Mi'kmaq. The explicit switch of audience is represented in the faces of the Mi'kmaw elder and children imploring, "Show me your math". The students respond to a community need to know rather than an external institution's need to know. In this revised interactional structure, the teacher can represent and mediate both external, mathematical knowledge and Mi'kmaw community knowledge.

In addition to the shift in audience, there is a shift in the construction of the student's identity. In the invitation, expressed both in the name of the contest and in the elder's and the children's call – "Show me *your* math!" – 'you' and the associated 'your' refer to the individual student, who is invited to address her or his community. Each individual is invited to work on his or her interests, not someone else's. By contrast, more typically in mathematics classrooms 'you' is used for generalizing – for example, "Your denominators must be equal when you add fractions". This generalizing sense of the word 'you' has been theorized by Rowland (2000) and exemplified by others, including Herbel-Eisenmann and Wagner (2007). The more personal 'you' and the presence of first-person pronouns – for example, the elder and children saying, "Show *me* your math," and the teenager saying, "*I* would like to find out more about …" – together with their recognition of person agency have the opposite effect of generalizing pronouns, which pervade mathematics. Morgan (1998) has noted the absence of personal pronouns as having a distancing effect in the relationship between students and mathematics.

However, Herbel-Eisenmann et al. (2010) have noted that in oral mathematics classroom discourse personal pronouns are more prevalent than in published mathematics resources. In fact, they are very prevalent in the most commonly used sets of words. In further analysis of these pervasive sets of words (called 'lexical bundles'), Herbel-Eisenmann and Wagner (2010) noted that there is still little room for personal latitude. Students are not invited to exercise their own choices very much. This research considered lower secondary-school mathematics classes, but we think the patterns extend into both elementary and upper secondary levels. In this research, Herbel-Eisenmann and Wagner describe a prevalent pattern in which students are positioned as doing things because their teacher 'wants them to'.

We argue that, in the case of the SMYM contest, the interactive pattern is similar in that students are responding to someone else's wishes, but that the situation is significantly different from typical classroom patterns because the students are responding to people in the community instead of their teachers and they are invited to make choices about what they want to study. Nevertheless, the SMYM contest is introduced by their teachers, so there is still the possibility of taking the positioning to be similar to typical mathematics classroom positioning.

A further complication in our representation of the positioning between students, their teachers and their community members relates to our lack of access to the many conversations that are associated with the SMYM contest. We as researchers do not have access to the conversations themselves, only to reports on these conversations, in the form of the student projects and other accounts of these conversations, arising from our conversations with community members, teachers and students after the key teacher–student conversations and student–community member conversations have taken place. For example, we do not know how teachers have mediated the video for getting the students going. However, even if they do not show students the video, it would have a structuring influence on the teacher's sense of the positioning being encouraged for the SMYM participation.

We also do not know details about the language of communication in the interactions between students and community members. We do not know the word choices and grammatical structures and we do not even know how many of these interactions were in Mi'kmaq. From our discussions with students and teachers, we know that the interactions would have been mostly in Mi'kmaq or in a hybrid of Mi'kmaq and English. In many communities, elders prefer to speak to children in Mi'kmaq and then occasionally translate into English. The elders would also have an expectation that the students make every effort to respond in Mi'kmaq. When English was spoken, it would have taken on grammatical structures of Mi'kmaq, which tends to be more verb-dominant and dynamic than noun-dominant and static (Lunney Borden 2010, 2011). The choice of language in these interactions is significant in terms of privileging community versus mathematical traditions. We believe that giving students the power to choose interactions that would privilege a language that connects to their identity, and that they do not use for mathematics, has the potential to transform mathematical understanding for them.

Our analysis of the positioning of students in the context of their SMYM contest work appears to contradict itself in various ways. There are elements of privileging colonialist or western control of mathematics and elements of release from this control. There are elements of teacher-mediated directions for students and elements of release from them. Altogether, this offers an example of the inherent complexity of identifying positioning in mathematics classrooms, one that is theorized further in Wagner and Herbel-Eisenmann (2009). Nevertheless, we feel that the SMYM contest continues to be worth doing because, even at its worst, it invites children to make choices about what mathematics they want to explore, connecting it to their community. However, in our view, the best justification for the SMYM project is that the larger Mi'kmaw community shows many signs of wanting it – and this is not a naïve community at all.

4.2 Challenges of Representation

We have described above how the children in the community engaged with their community representatives in relation to the SMYM contest and how we as researchers related to this. In addition to the positioning in the relationships within the community (including our work in the community), we all have positioned ourselves in relation to people outside the community. Students presented their findings on the internet (as stated above) and we as researchers have been reporting and continue to report to scholars and other educators on our conversations within the community.

Figure 5.3, which is an elaborated version of Fig. 5.2 which includes connections with people outside the community, represents communications from the SMYM conversations moving outside the community and the effects people outside can have on the people of the communities involved, both the children and adults. The box representing the people outside the community is shaded grey to indicate these people's agency. (The people outside the community affect us as researchers too, both directly and through our concern for our friends in the communities, but our focus for this chapter is on the situation faced by children who study mathematics in school.)

We note that the SMYM contest need not include the posting of student projects on the internet. There are a few good reasons for publishing the projects in this way. First, and foremost, many of the schools that have been active in the contest are part of a unique and relatively new jurisdictional agreement with Canada's federal government, giving the communities control over education. The communities are eager to demonstrate success within their schools to avert arguments for cancelling the agreement and are also anxious to make use of the technology provided by the First Nation Help Desk (http://firstnationhelp.com/) to ensure its continued funding. Positive publicity is in the communities' best interests. Second, many participant students have displayed their work at regional mathematics fairs, but most have not been able to attend. The internet provides a venue for all participants to share their discoveries with each other. Third, using internet posting aligns the contest with the other contests on which the SMYM contest is modelled, connecting it to the established storyline of community sharing.

In addition to the good reasons for publishing student projects in this way, however, there are concerns to consider. In our reporting on this research, the positioning theory lens helps identify some of them. We have experienced enthusiastic audiences in our reporting, but we have sometimes worried about the storylines that

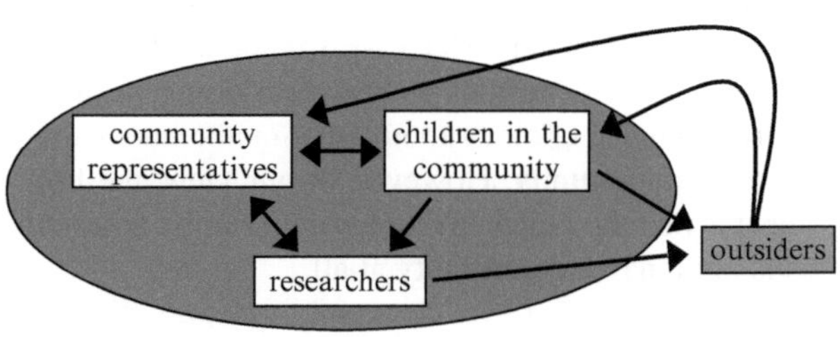

Fig 5.3 Extended interaction

might be enacted by our audiences. We have become aware of these storylines from the questions and feedback received by scholarly peers. For example, a colleague within the larger research project with which this research is associated wanted to use some of the students' ethnomathematical work to compose some problems for a textual resource he has been developing. From this, we were reminded that the students' ethnomathematical work could be used as a rich resource for people wanting to make connections between school mathematics and Aboriginal practices, and we also became aware of the lack of control the authors of the projects (the students) would have over how this material is used.

There are significant concerns to consider in representing Aboriginal community practices outside their communities. First, and most important (because our ethical responsibilities trump all other concerns in the research), we know that Aboriginal people in Canada are very concerned with the way they are represented outside their communities. Second, we share their concern ourselves and identify real dangers their communities face related to the images that feed stereotypes. There is the danger of essentialization. People reading a question taken from student ethnomathematical work may take it as representative of all Aboriginal communities or of all Aboriginal responses to the particular situation addressed. Aboriginal people, much to their detriment, have had and continue to have storylines attributed to their lives by outsiders in this way. In his Massey Lecture series, King (2003) has explained well (and satirized) the construction of the 'mythologized Indian' and some of the challenges such stereotyping presents for Aboriginal people.

Ironically, this problem of representation can be exacerbated by an emerging ethic of inclusion. School textbook publishers, clearly with good intentions, set standards for their books to include minimum percentages of representation of Aboriginal people in their images and word problems. The reality for authors and visual editors is that to meet these quotas they need to choose images and examples that are recognizably Aboriginal, which means using images and names that outsiders will connect with their knowledge of Aboriginal things, which inevitably includes stereotypes.

We looked through authorized mathematics textbooks for the Atlantic Provinces, which relate to the curriculum followed in the Mi'kmaw and Wolastoq[2] schools involved in the SMYM contest to get a sense of the current depiction of Aboriginal cultures in mathematics learning materials. We take all our examples from one book – *Mathematical Modeling – Book 1* (Barry et al. 2000), though we looked more widely. Our first observation was that there is very little in the books that situates the mathematics in any culture and plenty of missed opportunities. For example, with the question "What is the capacity of a pyramid-shaped box that is 20 cm tall and has a regular hexagonal base with side lengths of 15 cm?" (p. 269), the material for the box is dimensionless and students may wonder why someone would want a pyramid-shaped box. There are plenty of interesting-shaped containers in Aboriginal

[2]Wolastoq communities are often referred to as Maliseet. Wolastoqiyik and Mi'kmaq are neighbours geographically.

communities (and, for that matter, in non-Aboriginal communities), which have their own cultural distinctions worth including.

There are in this book some references to particular cultures, but they do not seem to honour the people of the culture. For example, off to one side in a box beside the main text (or perhaps emphasized, but nevertheless positioned separately from the main text) we found "Did you know? One of the most famous pyramids in Mexico is the Kukulcán located in Chichen Itza. The steps going up the pyramid are very steep." (p. 240) and a question in the regular text on the same page describing normal stairs: "For safety reasons, a 'normal' set of stairs can only have a rise of 72 cm for every 1 m of run. What is the tangent of the base angle B?" (p. 240). This combination positions the people of the Mayan Aboriginal culture as not normal and overlooks the significant design that went into constructing this pyramid by asking about a measurement that is relatively meaningless for someone building a pyramid. Why not ask questions that invite students to imagine themselves building such a pyramid and perhaps calculating how much stone they would need?

Gerofsky (2004) described the apparent arbitrariness of contextual and linguistic structure of word problems that implies an "'understanding' between writer and reader that these supposed situations do not have truth value, and that the writers' intentions and the readers' task are something other than to communicate and solve true problems" (p. 46). This apparent arbitrariness (and even disposability) of context may be particularly disturbing to cultures, such as Mi'kmaw culture, in which context is indispensable. Nevertheless, Gerofsky's discussions with mathematics students prompted her to note that word problems can give them a "point of entry, a place to insert oneself actively into the story" (p. 132). The point of entry could welcome the student's cultural knowledge. Alternatively, as seems to be the case with the positioning of Chichen Itza as not normal, the point of entry may engage students with antagonism by marginalizing non-European cultural traditions and thus alienate an Aboriginal student.

In mathematics resources, we would hope to see students being directed to imagine themselves in the shoes of someone doing mathematics to address their needs – an active point of entry. For this to happen, attention would need to be drawn to the questions one faces in design, instead of questions that one might ask about the finished product of the design. This distinction appeared in student contributions to the SMYM contest. It was clear that many of them positioned themselves alongside the designers in their community addressing personal and communal needs, as they identified explanations for how to make a *wi'kwam* (wigwam), a flat bread or a drum. Others, however, positioned themselves as an outsider using foreign mathematical tools to analyze a local product by, for example, using 'the formula' for the circumference of a circle to mathematize the outside of a dream-catcher.

Responding to concerns about representation, we recognize that if we did not report the ethnomathematical work done by the students in scholarship or in facilitating the students publishing their own work on the internet, people outside the community would still be positioning the community in certain ways. With publishing more positive and diverse examples and communications, there is hope that some stereotypes will be diminished. Most powerfully, there is a clear message that the members of the communities are exercising agency because the website that displays the projects attributes them to the students' work directly, because the diversity of the student projects suggests that they had significant liberty in their work, because the instructions are given by Mi'kmaw people in the video and because this is all displayed on a site hosted by a first nations (Aboriginal) organization – the First Nation Help Desk.

It is inevitable that there is a certain reciprocity in the relationship between the people in a community and outsiders. The members of the Mi'kmaw communities engaged in the SMYM contest have been speaking into the world of outsiders, who in turn speak into the community in various ways. We argue that repressing contact is not in the best interests of the communities. Rather, mindful consideration of the implications of engagement with outsiders is warranted and may draw attention to important mathematical values – for example, which is more valued, design or analysis? Nevertheless, when a community attends to the positioning at play within its relationships, as has been the case for participants in the SMYM contest and the people with whom they interacted, the view from the outside is more likely to be positive.

5 Reflection

As described above, our attention to interpersonal interaction illuminated aspects of our research activity. On reflection, trying to conceptualize the positioning with the maps given in Figs. 5.1, 5.2 and 5.3 illuminated even more. The map-making process and the maps themselves demonstrate to us that we have been seeing knowledge as a thing, as something that can be passed from one person to another. Such a conceptualization of knowledge can commodify it, as it established metaphors for the exchange and distribution of knowledge. Interestingly, seeing knowledge as a thing relates to the dominance of nouns in English speech and writing, relative to dominance of verbs in Mi'kmaq. The elders explanation of how to talk about a circle in Mi'kmaq, which we described briefly above, is an example of this difference.

Perhaps the context of conversations in an Aboriginal community further invited the commodification of knowledge because of well-known storylines that relate to 'keeping traditions', 'loss of language' and 'elders passing on their knowledge', all of which use nouns to refer to knowledge and tradition in the context of Western influence and use metaphors of possession and transactions. However, such metaphors are pervasive outside of Aboriginal communities too, where people talk about 'course delivery', 'acquisition of knowledge, attitudes and skills' and the 'possession of essential graduation learnings', among other images. We believe that the languages in the SMYM communities, Mi'kmaq and Wolastoqiyik, would not use these metaphors, because the languages are far less noun-intensive than English. This chapter depicts our metaphors because the accounts of our research comprise our reflections, not our Mi'kmaw counterparts' reflections. We look forward to talking with elders about this distinction.

Nevertheless, the most important reflection for us relates to equity, which we would characterize as the fundamental value that has driven the research we describe here and the other research we have done. In the above analysis, we analyzed the discourse in our research interaction as a means for addressing equity issues.

In Chap. 2, Gutiérrez clarified the scope of the word 'equity' as it has been used in mathematics education, distinguishing among access, achievement, identity and power. In our analysis, we have focused on identity and power, which Gutiérrez characterized as the critical axis of equity. Access and achievement form the dominant axis. She has elsewhere (Gutiérrez 2007b) described the critical axis as reflecting the mathematics that builds cultural identity around social and political issues, and notes that this kind of mathematics challenges static formalism. Our focus on the discourse, which highlighted interpersonal interaction, draws attention to the critical access. Nevertheless, we believe that such attention to power and identity in education relationships will have positive influences on achievement and access.

We understand that leaders in the Mi'kmaw communities do have an interest in access and achievement and we believe their interest in these aspects is not misplaced. In terms of access, they have articulated to us the need for people within their communities to be equipped to engage with external powers that greatly influence community concerns. For example, in the conversation in which the particulars of the contest were formed, community representatives' identification of modern community practices that should be highlighted in the ethnomathematics being done in the community comprised professions that would have immediate practical benefit for the community – plumbers, lawyers, lobbyists. It is recognized that for this kind of engagement mathematics is a key component. The shifts we made in our positioning as a result of paying attention to the discourse in our research relationships actually prompted school children to access community members who were engaged in professions that require mathematics. This access at a micro level is not what scholars usually mean when promoting or measuring access, but we think it is related. Students who have relationships that give them access to people in professional discourses end up making decisions that can give them access to these professions as participants.

It is inevitable that people position each other in their relationships. One way to avoid being positioned by others is to avoid relationships. We believe that there are greater dangers in isolation. We are suggesting that there is significant value for mathematics educators with an interest in equity to reflect on the discourses at play in their research. In particular, attention to interpersonal interactions was most fruitful in our experience, as it led to the fundamental restructuring of our research relationships and the resultant effects on the community context.

Asking questions such as the following may be a good way to begin. These are questions that were central to our criticism of the conversations in our research and to our choices for restructuring these conversations.

A teacher might ask:

1. *To whom are my students reporting their mathematics?*
2. *Whose problems/needs are my students addressing when they do the tasks I assign them?*
3. *How are people and communities represented in applications of mathematics I introduce?*

An education researcher might ask:

1. *To whom am I responding when I do my research?*
2. *Whose problems/needs are addressed in the research and how are these problems/needs identified?*
3. *How are people and communities represented in the reporting of my research?*

Chapter 6
How Equity Concerns Lead to Attention to Mathematical Discourse

Judit N. Moschkovich

In this chapter, I consider how equity issues are connected to mathematical discourse and what kinds of attention to discourse are relevant to equity. Using commentary on the preceding chapters, I discuss issues raised by different approaches to equity and to discourse. My first question about the two themes, equity and discourse, comes from asking how one would go about separating them. In looking at the two main sections in this book, I first wondered whose work belonged in which category – equity to discourse or discourse to equity – and, more importantly, how one would decide. I am not objecting to the distinction; in fact, in writing this chapter, I found it generative. I found myself thinking about the boundary between the two themes, not in opposition to the distinction but because I found the boundary interesting. In my own work, I cannot really make that distinction, because the connections between equity and discourse are dialectical (although it has been a struggle to maintain this two-way connection). It is possible that these two themes have been particularly connected for me, due to my personal history and intellectual trajectory.

I will explain with a little history, in order to locate the origins of my own interest in equity and discourse and also to provide a picture of my trajectory navigating the connections between them. I am originally from Argentina, the granddaughter of Jewish immigrants from Eastern Europe to Argentina and Brazil. My grandparents' first language was Yiddish, my mother's first language was Portuguese since she was born and raised in Brazil, and my father's first language is Spanish. My first language is Spanish, I learned some English in elementary school in Buenos Aires, but I did not consider myself bilingual until after I moved to the United States in high school.

Before becoming a researcher in mathematics education, I studied physics, mathematics and philosophy of science and taught mathematics at the college level for several years. I have worked principally in secondary mathematics classrooms

J.N. Moschkovich (✉)
Education Department, University of California at Santa Cruz, Santa Cruz, CA, USA
e-mail: jmoschko@ucsc.edu

B. Herbel-Eisenmann et al. (eds.), *Equity in Discourse for Mathematics Education: Theories, Practices, and Policies*, Mathematics Education Library 55,
DOI 10.1007/978-94-007-2813-4_6,

with students of Mexican, Puerto Rican and Central American origin living in the United States. I see my research as focusing on mathematical thinking, learning and discourse. My interest in bilingual learners blossomed a few years after finishing my dissertation in 1992. My personal experiences of learning a second language as a child, being an immigrant as a teenager and becoming bilingual as an adolescent sparked my curiosity about bilingualism and second language acquisition. My commitment to improving the education of learners who are from non-dominant groups provides the motivation and sustains my dedication to research. Overall, the perspective I bring to these issues is the sociocultural and situated one on language and bilingual learners that I have described elsewhere (Moschkovich 2002).

My Ph.D. work focused on cognitive science and mathematics education. I was able to address issues of what I then called "language" only in the last chapter in my dissertation – the rest of the thesis was a (very) cognitive analysis focused on describing student conceptions of linear functions and a second (more discourse-based) analysis of how these conceptions changed through discussions with a peer. This work was not presented or perceived as being focused on equity, even though all the students were from non-dominant, working-class communities (as well as bilingual in Spanish, Chinese or Tagalog) and one of the discussions I analyzed was bilingual in Spanish and English. During my post-doctoral positions (one at the Institute for Research on Learning, in Palo Alto, CA, and another at TERC, in Boston, MA), I was able to learn more about language, languages and discourse.

Although my initial work clearly focused on mathematics cognition and mathematical discourse, as soon as I started to work explicitly with Latino/a student populations and present analyses of bilingual mathematical discussions, I had an odd experience: my work started to be perceived as being about equity (when, in fact, this was the case even before my work focused on this specific student population). My experience is that those of us who study bilingual mathematical discussions are perceived as focusing on equity (not mathematical discourse, which is assumed to occur in monolingual mode), those of us who study cognition among students from non-dominant groups are perceived as studying equity (not mathematical thinking), while those of us who work in classrooms with immigrant children are perceived as focusing on equity (not learning and teaching in mathematics classrooms). Studying and working with a group of non-dominant students implies that the work we do is not about the things that human beings do: think, reason, talk or participate in mathematics classrooms, but about what one sub-set of human beings does. Therefore, this work is not about cognition, discourse or teaching writ large, but rather it is constructed as 'equity' work – in part, because it is relegated to being the study of how only *some* students learn (those students from non-dominant groups) and how we should teach those students.

This perception bothers me deeply for both practical and theoretical reasons. In terms of practice, this perception assumes that learners from non-dominant communities *are* the problem, because they learn in fundamentally different ways from regular folk, that teaching them requires special pedagogical tricks and that we cannot learn much about how regular folk learn (or how we should teach) from our work with learners from non-dominant communities.

In terms of theory, if the study of learning and teaching for learners from non-dominant groups is relegated to being only about that group, the study of learning and teaching (writ large) will continue to assume that there is a norm (regular folk, meaning those from dominant groups) and to reflect only the experiences of learners from dominant communities. Examples of so relegating non-dominant experiences abound in cognitive anthropology and cultural psychology, where the study of thinking by people from non-Western communities is not categorized as psychology (writ large) but as cross-cultural psychology (see Lave 1988 or Cole 1996, for a discussion).

Studies that use only participants from a dominant group assume these experiences to be representative of *human* thinking processes and set these experiences as a norm. Thus, a sample that is not representative is used to claim conclusions about the whole of human potential, as well as the range of variation in human thinking processes. One of my favorite examples of this phenomenon of using a select group to reach conclusions about general thinking processes is Perry's (1970/1999) work on intellectual development. Perry's studies were assumed to describe what his title did, namely 'intellectual and ethical development in the college years', yet were based only on data for Harvard and Radcliffe undergraduates. Although about 20% of the students were at least female, all the interviewees were part of a very select group of human beings.

Moreover, simply because work focuses on a non-dominant student population is not enough to claim that it address issues of equity. For example, I feel that my own work addresses equity issues not because I work with Latino/a students, but rather because I use a theoretical framework that focuses on the resources (and not the deficits) learners bring (and not the deficits) that teachers can use to support learning. In Vygotskian terms, I focus on *the potential for progress in what learners already know and do* (Vygotsky 1978). That shift in the analytical focus from deficits to resources reflects an epistemological stance toward knowing and power that fundamentally alters the analysis of where the power lies.

1 Multiple Approaches to Equity Discourse and Ethnomathematics

The chapters in this part of the book raise several issues regarding equity, discourse and ethnomathematics. I first summarize how each chapter addresses these issues and then consider each issue separately.

The first issue in connecting equity and discourse is how we approach (if not define) equity. In Chap. 2, Gutiérrez proposes four dimensions that are reflected in research addressing equity: access, achievement, identity and power. In her view, *access* relates to the tangible resources that students have available to them to participate in mathematics, including high-quality teaching, adequate technology and supplies, a rigorous curriculum, a classroom environment that invites participation, reasonable class sizes, tutoring, etc. *Achievement* focuses on tangible results for students at all levels of mathematics. Achievement involves course-taking patterns, standardized test scores and participation in mathematics courses at different academic levels (from elementary to graduate school). Studies focusing on identity examine whether students find mathematics meaningful to their lives and have opportunities to draw upon their cultural and linguistic resources (e.g. other languages and dialects, algorithms from other countries or different frames of reference). This dimension pays attention to whose perspectives and practices are valued. The *power* dimension can involve examining voice in the classroom – for example, who gets to talk and how contributions are taken up (or not).

Esmonde, in Chap. 4, defines equity as "fair distribution of opportunities to learn" (p. 52). She thus combines two of the dimensions suggested by Gutiérrez, namely access and power. Esmonde concerns herself with aspects of equity related to participation in classroom practices, focusing on how students take part, the impact this has on others' participation and also the effect of activity structures and local practices. She addresses an important question about access – access to what? – and considers student access not only to mathematical content and discourse practices, but also to positional identities. Esmonde considers positioning in terms of mathematical competence, which has a crucial connection to equity. This positioning works principally through discourse, in the sense of talk, dialogue or conversation among students and between students and the teacher (more on different senses of discourse later). In her study, Esmonde is careful to assert that identity and participation are not separate, but dialectically related. She also takes care not to equate participation in discourse practices only with talk. During group work, someone needs to be talking, but someone else also needs to be listening and thinking, as well as making meaning. And for a group's interactions to be equitable, the participation of each group member needs to be treated as important, not solely the contributions of one student positioned as smart or good at mathematics.

Esmonde explores two interesting situations that I think deserve further consideration. If each student in a small group is positioned as expert, then this is not equitable because even though they had all had equal opportunities to represent the group, none of them had opportunities to learn from each other. Since there were no opportunities for students to learn, the interactions were not equitable. Another situation should make us think further about how we structure the role of helper in small groups. She found that helpers began with their own understanding, rather than the understanding of the student who needed help. Thus, they developed a better sense of their own thinking, but did not help the student who needed help. In other words, 'the rich got richer' by working on their own ideas, rather than by listening to or building on the ideas of others.

Jorgensen describes issues central to school discourse, not only in mathematics but also in other content areas. At home, students learn not only the language of their communities (Spanish, English, etc.), but also specific varieties of that language appropriate to various social settings. In their home communities, they may or may not learn the language, dialect or register privileged in the school they will first attend. As important as the forms of language that children learn are the *uses* of language in their home communities and the ways people in various groups (e.g. children, adults, males, females) are expected to use it (Heath 1983, 1986). For example, language can be used for storytelling, for recounting experiences, for explaining natural phenomena and for entertaining.

Jorgensen describes how in some communities children are expected to respond to questions, are encouraged to ask questions or get praised for listening politely. These specifics point to the importance of knowing and honoring the discourse practices in students' communities of origin. As young children, students will have appropriated notions of what constitutes a story, how one talks about a past event, how one explains a task (or, rather, does not explain it but demonstrates how it is to be done) or how one engages in argument (Moschkovich and Nelson-Barber 2009).

The chapter by Wagner and Lunney Borden uses approaches to equity and to discourse that have both similarities and differences when compared with those of the two previous chapters. Their work on equity also addresses Gutiérrez' dimensions of access and power, but in ways different from Esmonde's research. In terms of access, Wagner and Lunney Borden are centrally concerned with supporting students from non-dominant groups to gain greater access to mathematics that is connected to their practices outside of school (at home, in the community, etc.), uncovering mathematics at work in local student communities and, overall, developing cultural sensitivity in mathematics teaching. Their hope is that culturally relevant curricula and resources will have an impact on the low participation in mathematics courses of Aboriginal peoples in Canada.

Discrepancies between students' own cultural practices and the cultural values of school mathematics instruction have been identified as a key reason for the lack of participation, interest and engagement in mathematics in Aboriginal student populations. Wagner and Lunney Borden recommend attention to differences in values and to appropriate teaching strategies. They also claim that an ethnomathematical approach to local mathematical practices "positions all mathematics as being culturally contingent" (p. 71) and their project draws on the expected outcome that "uncovered mathematical practices can inspire confidence in students who may assume they cannot do mathematics" (p. 72). This approach seems to parallel that of Jorgensen. However, she focuses not on the content of mathematics instruction, but on students' home language practices.

Wagner and Lunney Borden seem to define discourses as world-views: for example, when they refer to the "dominant discourses of the majority" (p. 71), "the discourses of mathematics and of cultures in conflict in colonialism" (p. 73) and when they refer to the differences between human interaction, in that interactions are "local, alive and dynamic" (p. 73). This is another way to define discourse, one that is broader than seeing discourse as text, talk, utterance, etc. I believe that this subtle difference in approaches to 'discourse' can be and has been confusing, so I will continue a bit further in contrasting different approaches to discourse, in the hope of clarifying what I have found confusing. I am not suggesting that this confusion is specific to these chapters or that it is in any way a defect. The confusion arises from the multiple interpretations and meanings of any word or concept. In my experience, these multiple interpretations cannot be avoided (and should probably be celebrated; see also Ryve 2011).

2 Three Points of Further Discussion

In this next section, I explore further three issues that arose for me in the preceding chapters: defining 'discourse', aspects of the discourse practices of school and challenges with the use of ethnomathematical approaches.

2.1 Defining 'discourse'

To start with, there is some confusion between the term 'language' and the term 'discourse'. Many commentaries on the role of academic language in teaching practice reduce the meaning of the term 'language' to single words and the proper use of grammar (for an example, see Cavanagh 2005). In contrast, work on the language of specific disciplines provides a more complex view of mathematical language (e.g. Pimm 1987), not only as specialized vocabulary (new words and new meanings for familiar words), but also as extended discourse that includes syntax and organization (Crowhurst 1994) and the mathematics register (Halliday 1978).

Theoretical positions in the research literature in mathematics education range from asserting that mathematics is a universal language through claiming that mathematics is a language to describing how mathematical language is a problem. Rather than joining in these arguments to consider whether mathematics is a language or reducing language to single words, I use a sociolinguistic framework to frame this chapter. From this theoretical perspective, language is a sociocultural, historical activity, not a thing that can either be mathematical or not, universal or not. From this perspective, the phrase 'the language of mathematics' does not mean a list of vocabulary words or grammar rules, but instead the communicative competence necessary and sufficient for successful participation in mathematical discourse.

One challenge in this endeavor arises because we all regularly participate in discourse and use language and, thus, we have developed intuitions about both discourse and language based on our personal experience. That experience with language is steeped in complex social, political and historical contexts and our intuitions may have developed into language attitudes. These may, at times, be in direct contradiction with empirical research on how people acquire language, use two languages or participate in conversations. Such intuitions lead to common pitfalls that need to be avoided when considering language and discourse in mathematics learning. One such is making superficial conclusions about language and cognition, such as that code-switching reflects forgetting a word or that the fact that a particular word does not exist in a national language means that speakers of that language cannot think of that concept. Both of these conclusions are massively contradicted by data.

There are multiple ways to approach and define discourse, from continuous text or utterances through Gee's conceptualization of Discourse (with a capital D) to the definition of discourse as world-view (for example, as in the phrase 'the discourse of colonialism'). I list only a few below (courtesy of the Merriam-Webster Dictionary), clustered into four categories:

1. (a) Connected speech or writing; (b) A linguistic unit (as a conversation or a story) larger than a sentence.
2. Verbal interchange of ideas, especially conversation.
3. Formal and orderly and usually extended expression of thought on a subject.
4. A mode of organizing knowledge, ideas or experience that is rooted in language and its concrete contexts (as history or institutions).

Gee's (1996) capital D 'Discourse' signals a view of discourse as more than just sequential speech or writing (#1), conversation (#2) or verbal presentation (#3). It may not (to my mind), however, quite extend to the broadest sense of discourse as world-view (#4). He writes:

> A Discourse is a socially accepted association among ways of using language, other symbolic expressions, and 'artifacts,' of thinking, feeling, believing, valuing and acting that can be used to identify oneself as a member of a socially meaningful group or 'social network,' or to signal (that one is playing) a socially meaningful role. (p. 131)

I would like to highlight some distinctions between the usual notion of discourse and Gee's definition of a Discourse. This is not the usual one used in linguistics textbooks, for instance specifying discourse as, "a sequence of sentences that 'go together' to constitute a unity, as in conversation, newspaper columns, stories, personal letters, and radio interviews" (Finegan and Besnier 1989, p. 526). Using Gee's definition, Discourses are more than sequential speech or writing and involve more than the use of technical language; they also involve points of view, communities, values and artifacts. Mathematical Discourses (in Gee's sense), then, would include more than ways of talking or writing; they would also include mathematical values, beliefs, points of view and artifacts. In particular, Gee (1999) reminds us to consider how 'stuff' other than language is relevant.

Overall, Gee's definition of Discourse provides a *situated* perspective and, in my opinion, has several advantages. It reminds us that *Discourse is more than language* in several ways, so that we can avoid that oversimplification. While Discourses certainly involve using language, they also involve other symbolic expressions, objects, people and communities. Another advantage of this definition is that it draws attention to the fact that *Discourses are situated both materially and socially*. Discourses involve not only talk, but also artifacts and social practices. And lastly, this definition assumes that *Discourses are not individual but collective*, or, as Hakuta and McLaughlin (1996) put it, "linguistic knowledge is situated not in the individual psyche but in a group's collective linguistic norms" (p. 608).

My own work (Moschkovich 2002, 2004, 2007a, b, c) is theoretically framed using a situated and sociocultural perspective on bilingual mathematics learners to identify the mathematical Discourse practices in student contributions (e.g. Moschkovich 1999). Following Gee (1996, 1999), I use the term 'Discourse' with a capital D to signal that I am using a situated view of discourse as more than utterance or text. Mathematical Discourse is not disembodied talk; talk is embedded in practices and these practices are tied to communities. I use the phrase 'mathematical

Discourse *practices*' (Moschkovich 2007b) instead of 'mathematical discourse' to highlight that Discourses are embedded in sociocultural practices, to emphasize the plurality of these practices and to connect Discourse to mathematical ideas.

Using the term 'practice'[1] shifts from purely cognitive accounts of mathematical activity to ones that presuppose the sociocultural nature of mathematical activity. I view mathematical Discourse practices as dialectically cognitive and social. On the one hand, mathematical Discourse practices are social, cultural and discursive because they arise from communities and mark membership in different Discourse communities. On the other, they are also cognitive, because they involve thinking, signs, tools and meanings. Mathematical Discourses are embedded in sociocultural practices. Words, utterances or texts have different meanings, functions and goals depending on the practices in which they are embedded. Mathematical Discourses occur in the context of practices and practices are tied to communities. Mathematical Discourse practices are constituted by actions, meanings for utterances, foci of attention and goals: these actions, meanings, foci and goals are embedded in practices.[2]

Talk is only one relevant semiotic system. Mathematical Discourse practices involve other symbolic expressions and objects. They involve multi-semiotic systems, not only speech, but also writing, images and gestures. The assumption that mathematical practices involve multi-semiotic systems is particularly important for analyzing mathematical activity cross-culturally. Otherwise, analysts might disregard semiotic systems (such as gestures and diagrams) that may be relevant.

Mathematical Discourse practices can also be connected to mathematical ideas. Cobb et al. (2001) define 'mathematical practices' as the "taken-as-shared ways of reasoning, arguing, and symbolizing established while discussing *particular* mathematical ideas" (p. 126). They contrast social norms and sociomathematical norms (which are not specific to any one mathematical idea) with mathematical practices (which, according to their definition, are). By focusing on mathematical Discourse practices that are specific to a particular mathematical idea, analyses can be grounded in particular mathematical concepts.

There is no one mathematical Discourse or practice (for a discussion of multiple mathematical Discourses, see Moschkovich 2002, 2007b). Mathematical Discourses involve different communities (e.g. mathematicians, teachers or students) and different genres (e.g. explanation, proof or presentation). Practices vary across communities of research mathematicians, traditional and reform classrooms. But even within each community there are practices that count as participation in competent mathematical

[1]I use the terms 'practice' and 'practices' in the sense of Scribner (1984), where a practice account of literacy serves to "highlight the culturally organized nature of significant literacy activities and their conceptual kinship to other culturally organized activities involving different technologies and symbol systems" (p. 13).

[2]For a description of how discourse practices involve actions and goals and an analysis of the role of goals in the appropriation of mathematical practices, see Moschkovich (2004). For an analysis of how meanings for utterances reflect particular ways to focus attention, see Moschkovich (2008).

Discourse. As Forman (1996) points out, particular qualities of argument, such as precision, brevity and logical coherence, are valued. In general, being precise, explicit, brief and logical, abstracting, generalizing and searching for certainty are highly valued activities in mathematical communities. For example, claims are applicable only to a precisely and explicitly defined set of situations as in the statement 'multiplication makes a number bigger, except when multiplying by a number smaller than 1'. Many times claims are also tied to mathematical representations such as graphs, tables or diagrams. Generalizing is also a valued practice, as in the statements 'the angles of any triangle add up to 180 degrees', 'parallel lines never meet' or '$a+b$ will always equal $b+a$'. Imagining (for example, infinity or zero), visualizing, hypothesizing and predicting are also valued Discourse practices.

2.2 *Issues with the Discourse Practices of School*

In her chapter, Jorgensen raises issues that are central to school discourse, not only in mathematics but also in other content areas. As part of school discourse, discourse practices in mathematics classrooms share some characteristics with school discourse in general. Although these characteristics are not specific to mathematics classrooms, they are central for children's success. As Jorgensen and other researchers have described, ways of organizing discourse can either include or exclude students from participating, and can have an impact on student achievement. Discourse practices in several communities – British working-class (Walkerdine 1988), native Hawaiian (e.g. Au 1980), Navaho (e.g. Vogt et al. 1987) and African American (e.g. Heath 1983; Lee 1993) – have been documented as being at odds with different practices in school. However, this and other research (Lee 1993; Lipka et al. 1998; Nelson-Barber and Lipka 2008) has also shown that it is not only possible, but also helpful, to incorporate children's own language practices into classrooms.

Teachers and researchers need to understand children's home language practices. This is not an either/or situation and Jorgensen is not suggesting replacing one set of practices with another. Teachers can learn to value and build on students' linguistic skills while also explicitly modeling the discourse styles expected in school. The rules about who can talk when, about what, in what ways and communication routines are established in every classroom. The practice of incorporating students' own ways of using language into the classroom is now recognized as one contributory aspect to the success of some classrooms. For example, one approach to integrating community language practices that resulted in gains in readings scores is the Kamehameha schools' integration of 'talk story' style of overlapping participation into native Hawaiian children's classrooms (Au 1980). Another example is Lee's (1993) work with African American high school students' ways of talking.

The question to ask about language practices in the classroom is whether a classroom facilitates comfortable and productive participation for students from non-dominant communities, in terms of the roles, responsibilities and styles of learners' communication practices. Answering this question means having substantial information about and deep understanding of children's home practices and the local community (Moschkovich and Nelson-Barber 2009). This entails not only knowing local activities that may be used in the mathematics classroom, but also students' language practices at home and in other community settings.

Such knowledge and understanding requires an ethnographic stance to research and practice. Such a stance draws on anthropology for notions of relativity that acknowledge the knowledge of the people studied (Spindler and Spindler 1997). A relativistic stance entails trying to understand the knowledge of others in their own terms as much as possible, *prior* to comparing it with other knowledge systems, including those of experts. Relativism allows us to move from deficiency models of learners to exploring their reasoning in terms of its potential for progress, a move that is especially relevant to research with learners from non-dominant communities. A relativistic stance towards culture avoids reducing cultural practices to essential or individual traits. Studying mathematical activity in context means not only considering the place where the activity occurs, but also considering how context – the *meaning* that the place and the practices have for the participants – is socially constructed. It is not sufficient to describe the setting in which learning takes place (classrooms, stores, homes); rather, reasoning and learning need to be described within the larger set of sociocultural practices that happen to occur in particular physical settings. In mathematics education, an ethnographic stance has been identified largely with ethnomathematics.

2.3 Issues with Using Ethnomathematical Approaches

Two of the chapters in this part of the book (by Jorgensen and by Wagner and Lunney Borden) illustrate several very important tensions in the use of ethnomathematical approaches. A central issue with these approaches, as Jorgensen points out, is that some ethnomathematical studies privilege western mathematics and fail to see the focal activity from the participants' perspectives. There are ways and examples of how to avoid this privileging. Two example of projects that seems to have managed to honor both local mathematical knowledge, as well as providing students access to the school mathematics they may need at other schooling situations, are the Yupik mathematics project (Lipka et al. 1998; Nelson-Barber and Lipka 2008) and the 'funds of knowledge' work (González et al. 2001).[3]

Uncovering the mathematics in any local activity involves making outside judgments as to what counts as mathematical. In some instances, this approach has been criticized, because it usually entails someone who is not a member of the local

[3]Work in mathematics seems to have focused on content more than language practices. Both of the works cited in the above paragraph focus on the content of instruction rather than on the language practices of the local community, by bringing into the classroom mathematical topics based on local activity. It is possible that some community language practices were also brought into classrooms. By knowing the student communities well, the researchers and teachers were most likely aware of language practices in the community.

community making that call from a position of power (a mathematician, a mathematics teacher, a researcher, etc.). On the one hand, this is problematic: in their chapter, Wagner and Lunney Borden provide examples of alternatives for finding and owning the mathematics in local activities. On the other, it seems that participants in everyday activity may regularly fail to see what is mathematical about what they are doing, unless it is arithmetic computation. For example, this was the case when I observed a group of insurance salespeople whom I could see were engaged in substantial mathematical activity, but when asked directly only reported doing anything mathematical when they used arithmetic. It is usually the work of the ethnographer and researcher to uncover the mathematical activity. This is a tension that ethnomathematical approaches may always confront.

While ethnomathematical approaches bring tensions with them, there are also several advantages to such approaches that make engaging with these tensions worthwhile. First, ethnomathematics provides an *ecological view of mathematical practices*, because it assumes that mathematical reasoning practices are multiple, heterogeneous and connected to other cultural practices. An ethnomathematical perspective is connected to equity issues that go beyond uncovering the mathematics in local activities to seeing children's mathematical competence in the classroom. Since this approach expands the kinds of activities considered mathematical beyond the mathematics in textbooks or schools (D'Ambrosio 1985; Nunes et al. 1993) and expands the definition of what counts as mathematical, we are more able to uncover and see the competence in learners' reasoning, even (and especially) when this reasoning may not look or sound like schooled mathematical thinking.

Using this perspective focuses data analysis on uncovering the mathematical structure in what participants are *actually* doing and saying. This kind of analysis makes students' mathematical activity more visible and describes the mathematical concepts students are grappling with, even when these concepts may not be immediately evident to participants or be expressed as formal mathematics. Taking an ethnomathematical stance means seeing student mathematical activity in the classroom not as a deviant or novice version of academic or school mathematical practices, but instead viewing it as an activity where participants use social and cognitive resources to make sense of situations.

In my own work, I use an ethnomathematical perspective to frame the description of mathematical activity among bilingual Latino/a learners. This framing is motivated by equity concerns and serves to avoid deficit models of learners. By shifting the focus from looking for deficits to recognizing the mathematics in student contributions, as well as by expanding the definition of what counts as mathematical, we are more able to uncover and see competence and thus avoid deficit models.

Lastly, Wagner and Lunney Borden raise a serious issue regarding how ethnomathematical approaches are perceived and used. If what we learn about the mathematical activities of any particular group or community is relegated to being only 'how those people do math', then we are in deep trouble in terms of equity. Here is an example of what I think the trouble is. When I tell people that I study how adolescents learn algebra, I usually get responses about bad experiences with algebra instruction. But when I tell someone that I work with Latino/a bilingual learners, the question I usually get is: *How do Latinos/as learn algebra*? I usually respond that they learn algebra in much the same way that all humans being learn mathematics – by making sense. But they do this in conditions that do not promote sense-making, such as instruction in a language they do not yet understand. My point is that underlying such questions is the assumption (supported by stereotypes) that Latino/a students learn differently from other adolescents, when instead the issue is that the *conditions* surrounding these learners are different. They are still a subset of human beings, a subset of adolescent learners, and so on.

These stereotyping issues may not go away until we uncover and address deeply held assumptions about intelligence or ability and how these notions are related to language, culture and particular communities. Another way to address these stereotypes is to be careful to distinguish between the *conditions of* learning and the *processes for* learning. For example, children in poor schools in the United States lack sufficient access to qualified teachers, advanced mathematics courses and material resources for learning (school buildings in decent condition, books, etc.). When we study learning and teaching mathematics in typical classrooms with students from non-dominant communities or look at their achievement scores on tests, we are thus reporting on the results of learning and teaching under the worst possible conditions, rather than on learning and teaching processes that would most benefit these students. Thus, it is important both to study and to disseminate examples of studies that describe teaching, curriculum, programs and approaches that have been successful for this student population.

3 Equitable and Successful Practices in U.S. Mathematics Classrooms

What might be equitable practices for students from non-dominant communities in mathematics classrooms? Overall, I would define equitable practices in mathematics classrooms as those that:

(a) support mathematical reasoning and mathematical discourse (because we know these lead to conceptual understanding and learning);
(b) broaden participation for students from non-dominant communities (because we know that participation is connected to reasoning and learning).

Classroom practices that support mathematical reasoning and broaden participation provide opportunities for students to use multiple semiotic resources to participate in, combine and value multiple mathematical discourse practices. Equitable classroom practices also honor student resources, in particular the 'repertoires of practice' among students from non-dominant communities.

Although research does not provide recipes for teaching, nor a quick fix, there are some general recommendations to guide researchers and teachers in developing their own approaches to supporting equitable practices in mathematics classrooms for students from non-dominant communities. For example, Brenner (1998) provides a framework for cultural relevance for instruction and curriculum (for more details, see also Moschkovich and Nelson-Barber 2009), one that includes the following considerations: *Do mathematical activities connect to those in the local community? Does the classroom facilitate comfortable and productive participation? Do roles and responsibilities fit with learners' communication practices? And does instruction enable children to build on their existing knowledge and experiences as resources?* Three of the chapters in this part show that these questions can be addressed in many different ways: Esmonde considers whether and how participation is productive for students; Wagner and Lunney Borden consider how mathematical activities connect to those in the local community; Jorgensen considers how classroom language practices fit with those of students' home communities.

Students from non-dominant communities also need access to curricula, instruction and teachers shown to be effective in supporting the academic *success* of these students. The general characteristics of such environments in the United States are that curricula provide "abundant and diverse opportunities for speaking, listening, reading, and writing [and that instruction] encourage[s] students to take risks, construct meaning, and seek reinterpretations of knowledge within compatible social contexts" (García and González 1995, p. 424). Some of the characteristics of teachers who have been documented as being successful with students from non-dominant communities are: (a) a high commitment to students' academic success and to student–home communication; (b) high expectations for all students; (c) the autonomy to change curriculum and instruction to meet the specific needs of students; (d) a rejection of models of their students as intellectually disadvantaged. Curriculum policies should follow the guidelines for traditionally underserved students (AERA 2006), such as instituting systems that broaden course-taking options, avoid systems of tracking students that limit their opportunities to learn and delay their exposure to college-preparatory mathematics coursework.

4 Some Recommendations for Future Research on Equity and Discourse

Looking back at the four chapters in this part of the book, I see agreement on the central directions for research that focuses on equity and/with/through discourse. In closing, I review how the four chapters serve as exemplars for future research in three ways (recommendations #1 to #3) and then, based on my own work (Moschkovich 2010), I make one further recommendation (#4).

4.1 Recommendation #1: Avoid Essentializing Cultural Practices

The four chapters use conceptual frameworks that do not essentialize cultural practices, nor describe culture as individual traits. In general, research should follow the examples set in this part and consider how students draw on multiple 'repertoires of practice' from home, everyday, school, etc. (Gutiérrez and Rogoff 2003). In order to avoid essentializing cultural practices, researchers suggest that we consider 'hybrid' practices (Gutiérrez et al. 2001) that are based on more than one language, dialect, register or practice.

In general, we can assume that communication styles and home cultural practices are heterogeneous and hybrid in any community, dominant or non-dominant (González 1995). Researchers working with populations of students from non-dominant communities should keep in mind that learners from any community can and do participate productively in a variety of roles, responsibilities, communication styles and mathematical activities that include hybrid practices. One example of a hybrid language practice is switching languages during a conversation, a practice called code-switching (for examples of code-switching work in mathematics, see Khisty 1995; Moschkovich 2002, 2007c; Setati 1998; Setati and Adler 2001). Monolingual speakers on both sides of national borders often perceive this practice as a deficiency. Regardless of what our personal experiences of code-switching may be, research in sociolinguistics (e.g. MacSwan 2000; Valdés-Fallis 1978, 1979; Zentella 1997) has shown that code-switching is a complex language practice that is not only cognitive but also social, cultural, historical and political, and, most importantly, not a deficiency.

4.2 *Recommendation #2: Avoid Deficit Models*

The four chapters also serve as exemplars of work that does not frame learners using deficit models. In different ways, using different approaches, the work in this part shows that there is a multitude of ways that research can avoid deficit models. All four chapters, in one way or another, focus on resources rather than deficits. This is a general way to avoid deficit models by considering not only the challenges students face, but also the resources (e.g., González et al. 2001) and competences (e.g. Moschkovich 2002, 2007a) they bring to mathematics classrooms.

Deficit models can be heard in comments that focus on what these learners cannot do, such as "These students cannot ______". While there is nothing inherently wrong with observing what students cannot do, deficit models are characterized by an emphasis on a lack of competence. If observations of what students *cannot* do are not accompanied by an analysis of what students *can* do, they provide an incomplete picture of these learners. Furthermore, deficit models of learners are usually tied to cause-and-effect explanations linked to the learners' home communities: these students cannot do x because their parents, homes or communities are not doing y. (For examples of how pervasive deficit models are, see McDermott and Varenne 1995.)

Two of the chapters (Jorgensen's and Wagner and Lunney Borden's) provide examples of how using a relativistic stance to study mathematical reasoning practices is a strategy for avoiding deficit models. This stance requires, in part, that when we observe learners from other cultural groups, linguistic communities or socioeconomic classes, we learn as much as possible about the norms of learners' home communities, not only through observation, but also by means of reading research studies that might be relevant. Empirical research on communication styles for non-dominant student populations may provide a relevant knowledge base for research. However, research on communication styles should be used with caution. These studies can serve as examples of how communication practices might vary, as Jorgensen does in her chapter, but not as a basis to make broad generalizations about the communication styles for any group of learners or any individual.

Another way to avoid deficit models is to move away from comparisons with a norm. For example, comparisons between bilingual and monolingual speakers are not a useful focus in mathematics classrooms (Moschkovich 2010), because they ignore competences that distinguish fluent bilinguals – such as code-switching – and miss how bilingual language competence is simply different from monolingual competence (Zentella 1997). Comparisons between monolingual and bilingual learners, students from dominant and non-dominant communities or speakers of standard English and speakers of other varieties, and so on, assume monolingualism, standard English or living where one was born as the norms for student experiences. Instead of focusing on comparisons to a norm that few students from non-dominant communities fit, studies need to examine student competences in their own right and explore the complexity of the experiences of students from non-dominant communities as they relate to mathematical reasoning, learning and instruction.

4.3 Recommendation #3: Recognize the Complexity of Language and Discourse Practices

There is also agreement across the four chapters in that the authors recognize the complexity of language and discourse, and have moved away from simplified views of language as vocabulary. Mathematical discourse is much more than vocabulary. While vocabulary is necessary, it is not sufficient. Learning to communicate mathematically is not merely or primarily a matter of learning vocabulary. The question is not whether students should learn vocabulary, but rather how instruction can best support students as they learn both vocabulary and mathematics. Vocabulary drill and practice is not the most effective instructional practice for learning either vocabulary or mathematics. Instead, vocabulary experts describe vocabulary acquisition as occurring most successfully in instructional contexts that are language-rich, actively involve students in using language, require both receptive and expressive understanding and require students to use words in multiple ways over extended periods of time (Blachowicz and Fisher 2000; Pressley 2000).

How is a complex view of mathematical discourse related to equity issues? The move away from discourse as vocabulary has crucial implications for equity. When mathematical discourse is reduced to vocabulary, students who come into classrooms from non-dominant communities are likely to be on the receiving end of this over-simplification. Their instruction will focus on superficial approaches to 'fixing' their lack of vocabulary. Instead of having opportunities to use mathematical language to communicate about and negotiate meaning for mathematical situations actively, their experiences will be reduced to the passive studying of vocabulary lists.

Another step in recognizing the complexity of language and discourse is to embrace the multi-modal and multi-semiotic nature of mathematical discourse (O'Halloran 2005; Radford et al. 2007): this move also has crucial implications for equity. Two issues to consider concern how participation is more than talk and how we interpret silences. Participation is more than talk: there is also quiet participation as evidenced by gaze, posture or later talk on the topic. If we assume that only students who talk are participating, we will miss the thoughtful yet engaged student who may be quiet and listening during a heated discussion, but joins in later with an insightful comment. It is also crucial to be careful about how we interpret silence. Although we can observe who does and does not speak, we cannot usually know why. Making inferences about imagined cultural, linguistic or cognitive reasons for silence is both unwarranted and dangerous.

Lastly, simplifying discourse usually involves creating dichotomies and these dichotomies often reflect differences in power. There are several dichotomies used to separate types of mathematical activity – for example, abstract/concrete or everyday/academic – that reflect a division of intellectual labor. These dichotomous categories are grounded on two fundamental assumptions: there are folk who have one and not the other; one of these is better or more valued than the other. Historically, those who have not had the kind of knowledge that is most valued have been those from non-dominant communities, colonized or marginalized not only by material conditions, but also by how researchers have perceived and labeled their thinking practices. Lave (1988) describes this division of intellectual labor:

> Functional theory underlies the web of relations between academic, novice, and jpf [just plain folk] worlds. In this theory, duality of the person translates into a division of (intellectual) labor between academics and "the rest" that puts primitive, lower class, (school) children's, female, and everyday thought in a single structural position *vis-à-vis* rational scientific thought. (p. 8)

This division of intellectual labor is fundamentally oppressive and inequitable. Therefore, shifting away from monolithic and dichotomous views of mathematical discourse practices is also closely tied to addressing equity issues.

4.4 Recommendation #4: Shift Away from Monolithic Views of Mathematical Discourse

Research and practice need to make a fundamental shift away first from conceiving *mathematical discourse* or *mathematical practices* as uniform and second from dichotomized views of discourse practices (such as 'everyday or academic'). Mathematical discourse is not a singular, monolithic or homogeneous discourse.

It is a system that includes multiple forms and ranges over a spectrum of mathematical discourse practices, such as academic, workplace, playground, street-selling, home, and so on. Researchers should consider the spectrum of mathematical activity as a continuum rather than reifying the separation between practices in out-of-school settings and the practices within school. Analyses should consider everyday and scientific discourses as interdependent, dialectical and related rather than assume they are mutually exclusive. Instead of debating whether an utterance, lesson or discussion does or does not count as being mathematical discourse, studies should instead explore what practices, inscriptions and talk mean to the participants and how they use them to accomplish their goals.

It is important for research and practice to move away from construing everyday and school mathematical practices as a dichotomous distinction. During mathematical discussions, students use multiple resources from their experiences across multiple settings, both in and out of school. Everyday practices should not be seen only as obstacles to participation in academic mathematical discourse. The origin of some mathematical discourse practices may be everyday practices and some aspects of everyday experiences can provide resources in the mathematics classroom.

Research needs to stop construing everyday and school mathematical practices as a dichotomous distinction for several reasons. First, a theoretical framing of everyday and academic practices (or spontaneous and scientific concepts) as dichotomous is not consistent with current interpretations of these Vygotskian constructs (e.g. O'Connor 1998; Vygotsky 1986). Vygotsky (and other theorists) describe everyday and academic practices as intertwined and dialectically connected. Second, because classroom discourse is a hybrid of academic and everyday discourses, multiple registers co-exist in mathematics classrooms. In general, Goody (1977) reminds us of the inadequacy of dichotomous categories for describing modes of thought or approaches to knowledge, "since both are present not only in the same societies but also in the same individuals" (p. 148).

Most importantly for supporting the success of students in classrooms, academic discourse needs to build on and link with the language students bring from their home communities. Therefore, everyday practices should not be seen as obstacles to participation in academic mathematical discourse, but as resources to build with, in order to engage students in the formal mathematical practices taught in classrooms. The ambiguity and multiplicity of meanings in everyday language should be recognized and treated not as a failure to be mathematically precise, but rather as fundamental to making sense of mathematical meanings and to learning mathematics with understanding.

Part II
Attention to Discourse Highlights Equity Concerns

Chapter 7
Linguistic Tools for Exploring Issues of Equity

Mary Schleppegrell

Research on the discourse of mathematics classrooms and curricula is making important contributions to our understanding of equity issues in mathematics teaching and learning. The language that presents mathematics to students communicates to them what mathematics is and the language teachers use in interacting with students positions them in particular ways as learners. This means that the linguistic choices that writers and speakers make in materials and interaction are consequential for the kind of learning that is offered and enabled. Language actively shapes our social world, as differences in wording construct different kinds of meanings and interaction. Exploring the language of teaching and learning mathematics and considering the kind of knowledge and opportunities for participation afforded by different ways of using language can therefore offer researchers important insights into issues of equity.

This chapter identifies equity issues that can be explored through analysis of language and reviews studies that have applied linguistic tools to inform those issues, using Gutiérrez' (2002a, 2007b) framework to situate the linguistic analyses in terms of their contributions to our understanding of equity. Gutiérrez suggests that equity of *access* and *achievement* enables students to learn the dominant mathematics. This chapter addresses these equity issues by describing linguistic tools that enable researchers to explore the nature of the mathematics being made available to students, as well as the nature of the mathematics knowledge that students appropriate and develop as they learn. Gutiérrez also suggests that equity of *power* and *identity* enables students to challenge the dominant mathematics. This chapter addresses these equity issues by describing linguistic tools that enable researchers to explore how students are positioned in mathematics classrooms and whether mathematics curricula open up opportunities for students to experience themselves as

M. Schleppegrell (✉)
School of Education, University of Michigan, Ann Arbor, MI, USA
e-mail: mjschlep@umich.edu

B. Herbel-Eisenmann et al. (eds.), *Equity in Discourse for Mathematics Education: Theories, Practices, and Policies*, Mathematics Education Library 55,
DOI 10.1007/978-94-007-2813-4_7,

participants in shaping the mathematics they are learning. In her analyses of successful U.S. high school mathematics departments, mathematics teachers and mathematics teacher candidates, Gutiérrez highlights the importance of linguistic practices, showing how equity is discursively constructed and how some forms of discourse provide more equitable conditions for learning in particular contexts than others. She argues for closer attention to discourse to recognize how it can support students' equitable participation in mathematics learning.

This chapter reviews studies that have used linguistic tools to explore equity issues in mathematics teaching and learning, showing how close attention to the forms language takes can inform key questions that mathematics education researchers have asked. The linguistic tools described in this chapter come primarily from systemic functional linguistics[1] (SFL), a theory of language that recognizes that language varies according to the context of use (Halliday 1978; Halliday and Matthiessen 2004). It is a socially engaged theory that offers tools for making connections between the language used by speakers and writers and the equity issues that researchers are concerned with, as it offers a fully-developed discourse grammar that can be used to analyze the language of pedagogical materials, classroom interaction, students' written products and other uses of language (in connection with mathematics symbolic language and visual display) in the mathematics classroom.

A key notion in SFL is that language is a powerful means of construing our social reality and of enacting social relationships. Speakers and writers are constantly making choices as they use language. These are not always or even mainly conscious choices, but are nevertheless particular wording selections that can be contrasted with other possible ways of wording. The choices come out of the contexts in which speakers interact and writers envision their readers and these choices in turn construe those contexts in particular ways. In classrooms, teachers and materials are always simultaneously presenting knowledge and positioning students in relation to that knowledge, directing their attention and actions and evaluating their performance (Christie 2002). Analysis using linguistic tools helps us see how mathematics is presented to students, how learning is accomplished in mathematics classroom discourse and how different kinds of participation result in different kinds of learning. These tools help us explore questions related to equity such as the following:

- *What is the nature of the mathematics being offered to students through classroom discourse and pedagogical materials? What views of mathematics and of mathematical activity do students construct as they participate in learning?*

These questions address equity issues related to the nature of the mathematics being offered to and taken up by students, illuminating issues of access and achievement. They can be explored by analyzing the *thematic patterns* in the ways mathematics

[1] See Schleppegrell (2004) for an overview. For a complementary discussion of SFL and arguments and examples related to its use in mathematics discourse analysis, see Morgan (2006). O'Halloran (2005) provides a detailed analysis of mathematics discourse using SFL. For a more extensive review of research on language in mathematics, see Schleppegrell (2010).

is construed by teachers, students and texts and by exploring the *process–participant configurations* and *modality* in the language used in mathematics classrooms.

- *What are the processes through which knowledge is developed? How are students positioned as learners through classroom and pedagogical discourses? How does teacher interaction with students mediate their learning?*

These questions address whether students are construed as powerful in participation and as contributors to the evolution of knowledge in mathematics, illuminating issues of power and identity. Exploring them helps reveal how a teacher mediates students' developing knowledge about mathematics. Linguistic approaches such as analysis of *process–participant configurations* can identify the ways students are positioned by the discourses they engage in, while analysis of *modality* and *mood/ speech function* reveals how students' actions are regulated by the teacher as they learn mathematics.

1 Analyzing the Mathematics Being Presented Through Thematic Patterns

Achieving equity of *access* and *achievement* depends on offering students instruction that maintains the integrity of the mathematics itself and that develops students' understanding of that mathematics over time. Investigating the ways mathematics concepts are presented to students in materials and interaction, and exploring how students themselves represent the mathematics they are learning, are both ways of examining what students are being provided with access to and how they are achieving. One linguistic approach to exploring these issues is analysis of *thematic patterns.* Thematic patterns are the relationships between the constructs being learned that are built up by teachers and students in interaction or are presented in curriculum materials as particular topics are developed. Analysis of thematic patterns offers a method for identifying key concepts and seeing how they are presented to students and how students take them up, providing a powerful tool for investigating the knowledge that is being made available to students and the learning that is taking place. It can answer questions such as: *Is the mathematics that students are being offered accurate, appropriate and challenging? Does it connect with their prior knowledge and help them move toward more complex understandings?*

The study of thematic patterns in a U.S. secondary science classroom is illustrated by Lemke (1990), who analyzes what a physics teacher says to students and how students' language represents the same knowledge. Lemke demonstrates that the teacher's and students' discourses represent different understandings of the relationship between heat energy and light energy. He argues that when students use a different thematic pattern from the one the teacher does in talking about a topic, they are also thinking about it in different ways. Lemke characterizes mastery of thematic patterns as the most essential element in learning and presents a method for studying these patterns.

Chapman (1995, 2003) illustrates application of this approach in her study of the discourse through which the notion of *function* is developed over the course of a unit of study in a year 9 mathematics classroom in Western Australia. She uses analysis of thematic patterns to show how a teacher develops students' understanding of the notion of a *constant* (or *common*) *difference* in finding the slope of a line. To identify a thematic pattern, Chapman analyzes the ways mathematics constructs are presented in language and how they are related to each other through close attention to the wording used by the teacher, the mathematics materials and the students. This enables her to demonstrate how the representation and understanding of a *constant* (or *common*) *difference* evolves over the course of the unit.

For example, on the first day of the unit, the teacher and students are discussing a homework problem that presented a growing sequence of dots. The students were asked to calculate differences between the number of dots in each set. Here is how the teacher begins (emphasis and ellipsis added):

Teacher: …is there anyone who got *a rule*? (nominates a student, Ryan, to go to the board and demonstrate and references the textbook problem). Ryan, just go straight to *the numbers. X and y table.*

(Chapman 1995, p. 246)

At this point, the thematic pattern that the teacher has presented can be recognized in the ways *a rule, the numbers* and an *x and y table* are put into relationship with one another. This is represented in Fig. 7.1.

Notice that the teacher's first utterance suggests that the *rule* will come straight from *the numbers.* The teacher then expands this understanding through interaction with Ryan:

Teacher: For those who are struggling to find what's going on, we are looking at *a pattern of dots* which generated *these numbers.* … OK. What did you do?

Ryan: I tried to put in, ummm.

Teacher: Well, you just put those little *n*'s in so I can see where *the difference pattern* comes between. All right?

Ryan: Then I just kept working it down and then that was to the–

Teacher: Just doing it one more time so those people who really believe it, did you all find this? Did you run through this just one time, hoping to get somewhere, looking for the common difference and all that looking for *the common difference* … keep working it down. Does that mean *the difference pattern* is in actual fact two?"

Student: No, because it's not constant. You can't tell if it's constant or not because there's only one number.

(p. 246)

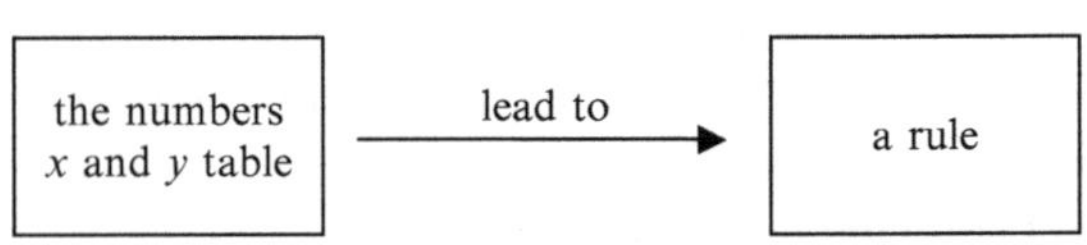

Fig. 7.1 First thematic pattern (Adapted from Chapman 1995)

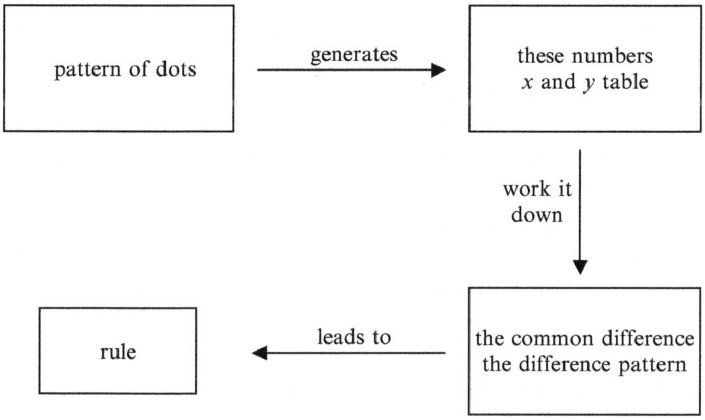

Fig. 7.2 Second thematic pattern (Adapted from Chapman 1995)

Over the course of this interaction, the procedure for solving the problem is further developed, representing it as a process of *working it down* to find the *common difference* or *difference pattern*. Figure 7.2 presents the thematic patterns that have been developed at this point in the lesson, showing the complexity that has now been added to students' understanding about how to find the *rule*. Chapman continues to trace the ways semantic relationships are developed as students read mathematics texts and discuss the concepts they are learning, showing how meanings related to functions are greatly elaborated over the course of instruction.

To create visual representations of the thematic patterns emerging from this talk, Chapman tracks the evolution of the key mathematical concepts by identifying the grammatical *participants* (nouns and noun phrases italicized in the excerpts above) and the relationships between them that are constructed in the discourse through the grammatical *processes* they are involved in (verbs and verb phrases underlined in the excerpts above). Identifying grammatical processes and participants and creating schematic representations of thematic patterns offer a means of exploring how mathematical meaning is built up over time in classroom discourse. It has the potential to illuminate aspects of the mathematics that students are presented with, as well as providing a means to chart their taking up of mathematics concepts and to trace where their understanding is different from that of the teacher or the textbook.

Thematic pattern analysis provides evidence of the integrity of the mathematics that is construed in the classroom and thus can be used to illuminate equity issues related to access and achievement. Through analysis of the language of the textbook materials, for example, the ways mathematics concepts are presented in the official curriculum can be explored and thematic patterns developed in different kinds of classrooms and different contexts of learning can be compared, in order to look closely at the nature of the mathematics that is being offered to students (for an example of this, see Herbel-Eisenmann and Otten 2011). This helps us recognize whether or not students in different contexts have equal access to understanding mathematical meanings.

By analyzing the thematic patterns in the language used by students to talk about mathematics, we can see how mathematical relationships may be taken up in different ways by learners. (See also O'Halloran 1998, for analysis of thematic patterns in mathematics texts and oral discourse.) In addition, the thematic patterns in the mathematics students bring from their home contexts can be compared with the thematic patterns of the 'official' mathematics in order to recognize how mathematics itself can be construed in various ways. The analysis of thematic patterns can assist in evaluating equity in access and achievement by revealing the mathematics represented in the discourse of teachers, students and materials.

2 Exploring the Positioning of Students Through Process/Participant Analysis

Not only can a focus on grammatical processes and participants reveal the nature of the mathematics offered to students, it can also help researchers explore the ways students are positioned as learners and contributors to mathematics, both in the curricular discourse and in classroom interaction. A focus on the different kinds of grammatical processes and participants represented in mathematics discourse helps the researcher explore the ways discourse positions students and directs their actions and thinking, thereby illuminating issues of access, power and identity. How students are positioned as learners affects the power they feel as participants in the mathematics classroom and the identities they are constructing for themselves as mathematics learners. Analysis of the grammatical processes and participants in mathematics discourse can help researchers explore questions such as: *Is the mathematics being presented construed as something procedural or as something that requires thinking and analysis? Are students positioned as potential participants in the development of mathematics knowledge? Are they treated as thinkers as well as procedural actors?* These are issues with clear implications for equity, as different ways of construing mathematics and positioning learners have consequences for students' access and achievement, as well as their sense of power and identity.

From a functional linguistics perspective, processes can be categorized into different types, based on their meanings and the grammatical configurations they present.[2] Both instructional materials and classroom discourse can be analyzed to identify the kinds of processes that are prominent and what they are accomplishing for the speaker/writer, in order to recognize how pedagogical materials represent mathematics and how those "official" representations are taken up in the classroom (e.g. Herbel-Eisenmann 2007; Herbel-Eisenmann and Wagner 2005; Morgan 2005). For example, researchers can explore who is acting and thinking at different points in the discourse or who is

[2]Different SFL analysts divide the experiential space in different ways, with Halliday and Matthiessen (2004) offering categories of *material, behavioral, verbal, mental, relational* and *existential* meanings, while Martin and Rose (2003) conceptualize processes as of four types, *doing, saying, sensing* and *being*. As with all linguistic notions, these categories can be specified in greater or lesser detail, depending on the goals of the analysis.

being directed to act or think in some way, by identifying and comparing the use of *material* and *mental* processes and the participants in them. *Material* processes are processes that are 'external'; *doing* processes that describe actions or happenings in the world. *Mental* processes are 'internal'; processes of *thinking, perceiving, feeling and experiencing* in human consciousness. Researchers can examine *relational* processes, processes of *defining, describing, having* and *being*, to explore what is defined and described. Each of these different process types has its own set of participant roles:

Mental process:

We *know* that there is a procedure for finding the answer.
Senser **Phenomenon**

Relational process (attributive):

The triangle *has* a perimeter of 36".
Carrier **Attribute**

Relational process (identifying):

A triangle *is* a three-sided figure.
Token **Value**

Material process:

The student *added* the numbers.
Actor **Goal**

Note that the semantic roles in a material process do not change, even when passive voice is used:

Material process:

The numbers *were added* by the student.
Goal **Actor**

If the clause does not say who the actor is, then it is referred to as an agentless clause:

Material process:

The numbers *were added.*
Goal

A full discussion of participants in these different process types is beyond the scope of this chapter – material processes, for example, may also have Beneficiaries, Ranges and other participant roles (see Halliday and Matthiessen 2004).[3]

Looking at different process types and the participants in them, researchers can explore the nature of students' positioning in relation to the mathematics; who is construed as agentive in doing mathematics and whether the discourse includes or

[3]The key point is that different process types are distinguished on the basis of grammatical criteria; for further information, see Halliday and Matthiessen (2004) or Fang and Schleppegrell (2008).

excludes the student as a knower/contributor to mathematics. For example, Morgan (2005) analyzes mathematics research papers and students' textbooks in the U.K. She investigates how the authors present definitions by looking at relationships between process and participants in material processes to focus attention on how agency is represented (material processes in italics):

> *From a mathematics research paper*: we *give* a somewhat non-standard definition [...] This viewpoint *makes* the actions [...] more or less transparent (p. 110)
>
> *From a GCSE school text*: you *found* that the ratio [...] This ratio *is given* a special name. It is called ...(p. 112)

Morgan points out that the mathematics research paper constructs explicit human agency in identifying *who* has developed the definition: *we.* The student text, on the other hand, obscures the agency involved in defining through use of the passive voice; we are not told *who* gives the ratio a special name. The analysis of the relationships between processes and participants also yields other interesting findings here. Morgan observes that, after presenting the definition, the authors of the mathematics research paper use the nominalization *this viewpoint* (a nominalization is the linguistic construal of a process as a thing; here, the *giving of the somewhat non-standard definition* becomes *this viewpoint*). The *viewpoint* is then an agent in the process *makes the actions transparent;* in other words, this abstraction is presented as an actor, performing the process of *making the action more or less transparent.* Through this analysis, Morgan illuminates the ways these texts represent the nature of mathematics and construe power and authority in different ways.

Morgan (2006) further develops this perspective as she explores how learners are positioned as they engage with mathematical knowledge. Analyzing secondary school textbooks and writing produced by secondary school students in England on GCSE exams, she compares what is presented in material and relational processes to show that in some cases the language of mathematics constructs students as actively *doing* while in other texts the mathematics is just *given* (e.g. the difference between *if you do this, X increases* and *here is the formula*). These different ways of positioning the students are not neutral. Morgan suggests that when textbooks use procedural discourse as an organizational strategy, constructing the students as actors who are to perform particular actions, they "may make students more likely to perceive mathematics as consisting of a set of procedures and hence, perhaps, to find it more difficult to engage with relational or logical aspects of the subject" (Morgan 2006, p. 228). Texts that obscure human agency, on the other hand, "may contribute to difficulties for some students in seeing themselves as potential mathematicians" (p. 228).

Morgan compares texts written by two students in response to a problem, looking at how students represent mathematical objects, the processes they are involved in and who is acting in the processes. By analyzing the patterns constructed by the students, she is able to show that one student draws "primarily on a discourse of investigation, oriented to value exploration of interesting mathematics", while the other "draws strongly on an assessment discourse, displaying the 'answers' valued within that discourse" (p. 236). Morgan argues that differences in wording "construct different images of the objects of mathematics and the nature of mathematical activity. At the same time they claim different types of authority and construct different 'ideal' positions for their readers" (p. 236). Analysis of processes and participants in the grammar enables researchers interested in issues of equity to reveal how students and mathematics are positioned and how students themselves represent the mathematics they are learning.

Herbel-Eisenmann and Schleppegrell (2008) use the notion of processes of *doing* (material processes) and processes of *thinking* (mental processes) in showing how a teacher in a U.S. middle school classroom mediates students' learning and directs their attention and actions by explicitly linking action with reflection in her classroom discourse. They point out that, when the teacher asks *What question would I be asking myself in my head as I start that problem?* (p. 29), the teacher is developing in students the notion that the solving of the problem is also engaging them in a thinking process and that this linking of reflection and action is part of what mathematics is about. This, along with other key moves that the teacher makes, positions the students as mathematical *thinkers* and positions mathematics as something that is about meaning, has reasons and requires engagement in reasoning and explaining. (See also González 2011, for examples of this kind of analysis related to what geometry teachers expect students to be responsible for.)

The notion of processes of different types has also been used in mathematics research that draws on other analytic frameworks. An example compatible with but not drawn from SFL is Rotman's (2000) distinction between 'exclusive' ('action'-oriented) verbs, describing actions that can be done independently from others, and 'inclusive' ('thinking'-oriented) verbs, requiring dialogue. As 'inclusive' commands in mathematics he gives the examples *consider, define* and *prove* and their synonyms and suggests that other commands are 'exclusive' (p. 10). This distinction has been adopted by mathematics discourse analysts to explore how students are positioned (e.g. Burton and Morgan 2000; Herbel-Eisenmann and Wagner 2005).

Wagner and Herbel-Eisenmann (2007), for example, analyze the processes that typically follow the word *just* in mathematics classroom discourse, finding Rotman's distinction productive in thinking about the contexts in which this common expression ('just') occurs in mathematics and what it means for students' engagement. They show how the processes that follow the word *just* are typically 'exclusive' in Rotman's terms and argue that this constructs a monoglossic rather than dialogic discourse. Their analysis illustrates how dialogue can either be closed down or opened up by the teacher's language choices. See also Mesa and Chang (2010) for analysis of these same issues using the SFL constructs *appraisal* and *engagement* (Martin and White 2005).

Pronouns have been a focus in the analysis of mathematics discourse since Pimm's (1987) classic study, which built on Halliday (1978). Rowland (1999) explores the meanings construed in pronouns, suggesting that analysis of how *we* is used constructs relationships that are either inclusive (*we* meaning *you and me*) or exclusive (*we* meaning *my co-authors and I*) and that this can reveal who is being construed as involved in mathematical thinking and doing. The pronoun *you* has also been a focus of study, as it may construe a generalized reader/listener or a specific reader/listener who is then instructed to *do* or *think* in some way. Herbel-Eisenmann (2007), for example, finds that commonly-used U.S. middle school curriculum materials often actually tell students what *you found* in a previous inquiry as they begin a new investigation. She suggests that this constructs a "common readership that had done and found certain things" (p. 14) – one perhaps related to the author's need to depend on certain presumed/shared knowledge in order to move forward. From the point of view of the reader, however, she points out that such language may be read as an attempt to control and define the students' own sense-making.

3 Exploring Contingency and Authority in Mathematics Through Analysis of Modality

Researchers can also analyze mathematics discourse to explore how learners in particular social contexts may be experiencing mathematics or classroom interaction in ways that give them less access or power or in ways that affect their achievement or identities. A linguistic construct that is often used for this purpose is *modality*. Modality is the linguistic system used to mark degrees of *possibility, usuality, obligation, inclination and ability*. Analysis of the modality of *possibility* (what *may* or *might* be) helps us explore the strength of claims being made and whether knowledge is being presented as contingent or absolute. Analysis of the modality of *obligation* (what *must, should,* or *ought to* be done) helps us explore what is being construed as necessary and, consequently, how learners are being regulated. Analysis of the modality of *ability* (what *can* or *could* be) helps us look at how learners are being positioned as competent to engage in particular activities. Modality can be realized in modal verbs (e.g. *can, could, may, might, must, ought, should*, etc.), as well as in other language structures such as adverbs (e.g. *maybe, usually*).

Use of modality contributes to enacting a particular kind of relationship between writer and reader or speaker and listener, enabling degrees of contingency or obligation to be construed. Modality affects how the mathematics itself is positioned; whether as something contingent and evolving or as something fixed and authoritative. Use of modality also has implications for how interactants are positioning each other and the degree of control that is being exercised. By analyzing the meanings construed through modality, researchers explore issues of confidence, ambiguity and authority (e.g. Burton and Morgan 2000).

Morgan (2005, p. 112), for example, uses analysis of modality along with the analysis of processes and participants described above to compare the ways definitions are presented in published research and in GSCE Intermediate and Higher textbooks in the U.K. (my ellipsis and italics):

Intermediate text: you found that the ratio [...] is the same for each of these triangles. *This ratio is given a special name.* It is called ...

Higher text: The ratios [...] *may be defined* in relation to the lengths of the sides of a right-angled triangle ...

As Morgan points out, both texts use passive voice and obscure who is responsible for these definitions, but "[w]hile the Intermediate text lays down a set of absolute and unquestionable facts to be accepted by the student-reader, the Higher text allows uncertainty and alternatives, opening up the possibility that the student-reader herself might choose between the two definitions" (p. 114).

Morgan's analysis of modality across a range of texts enables her to argue that in academic mathematics and higher-level textbooks, definitions are presented as dynamic and evolving, open to decision-making by the mathematician and incorporating ambiguity, while lower-level textbooks typically present definitions as static. She shows how professional mathematics texts are very different from school texts in the kind of reader they construct and that pedagogical texts aimed at different levels of students also vary in the positioning of the reader as more or less invited into or involved in the decision-making activity that the text constructs. The texts also construe mathematics in different ways, as relatively more evolving (the professional texts) versus given and absolute (the pedagogical texts).

Morgan points out that ambiguity and multiplicity of meanings are important for learning and that the pedagogical construal of mathematics as defined and lacking human agency does not engage students in the kind of mathematical meaning-making that the professional discourses of mathematics offer. She suggests that allowing for ambiguity and multiplicity of meanings in classroom discussion is important for supporting students' developing understanding. Morgan also raises questions about the focus on vocabulary and defining terms that is common in pedagogical practice and is promoted in mathematics policy documents such as the U.K. National Numeracy Strategy. She suggests that 'clear explanations' may not always be the most important goal in teaching, pointing to the more advanced texts that provide models for students of creativity and purposefulness in mathematical practice.

Another example of discourse analysis that focuses on the ways mathematics materials position learners is Herbel-Eisenmann (2007), who looks at these issues in terms of the 'voice' constructed in a U.S. grade eight school mathematics unit based on reform principles that are intended to shift the locus of authority in mathematics toward the student. She operationalizes *voice* through analysis of several features, including processes, pronouns and modality. The materials she describes generally do not draw on the modality of *possibility* except to talk about what results a reader *may* or *might* have found. On the other hand, she finds many examples of *obligation* and *inclination* (what students *should, would* or *will* do). She describes a general high degree of absolute modality and argues that the ideological goal of the intended curriculum – that the reader should be positioned as someone who can engage in classroom discussion about mathematics concepts rather than as someone who should take the textbook and teacher as the only sources of authority – is not easily realized, raising questions about the ability of conventional textbook forms to present students as more authoritative readers.

4 Exploring Classroom Interaction Through Analysis of Mood and Speech Function

It is in the process of interaction in the classroom that students' sense of themselves as engaged in mathematics and students' sense of the intellectual significance of mathematics develop. Exploring how teachers construct mathematical understanding in interaction with students can help us think more deeply about equity issues in classrooms with diverse learners, where students' backgrounds and experiences may position them to engage in different ways. Analysis of the meditating role of the teacher can also reveal differences in opportunities students are being given to engage with mathematics or differences in the ways teachers interact with students in different contexts and settings.

Linguistic constructs that are useful for analysis of classroom interaction include *mood* and *speech function. Mood* is a general linguistics construct, referring to the three possibilities for structuring a clause in terms of interaction: *declarative, interrogative and imperative. Speech function* refers to the meaning made by the mood choice, which is not always congruent with the grammatical form, as different moods can construe the same speech function, creating variation in discourse that is related to power and authority, as well as cultural conventions and practices. For example, to request that a student do something, a teacher may use any of the three grammatical moods, saying *Please take out your books* (imperative); *Would you take out your books?* (interrogative); *I'm looking around for people who have their books out* (declarative).

Table 7.1 shows how these constructs interact, illustrating the 'typical' mood used for particular speech functions and the variability that is possible. The interaction of mood and speech function allows flexibility and variability in the way teachers direct students, but sometimes 'mismatches' between mood and speech function lead to misunderstanding, especially where teacher and students do not

Table 7.1 Typical grammatical forms of the speech functions

Speech function	Mood
Statement	Typical: Declarative: *I gave you this problem already.*
Question	Typical: Interrogative: *What's the answer to number two?*
	But also sometimes declarative: *I'm wondering if anyone has the answer to number two.*
	Or imperative: *Raise your hand if you have the answer to number two.*
Command	Typical: Imperative: *Look here!*
	But also sometimes interrogative: *Can I have your attention?*
	Or declarative: *I'm waiting to see that you are ready.*
Offer/Request	No typical mood.
	Let's have another person come to the board now.
	Would you like to do the second problem or number three?
	How about starting that again now.
	Go ahead and choose the next one. (etc.)

share the same cultural ways of questioning and commanding. This can be seen in utterances such as *Would you like to do the second problem?*, where the grammatical mood is interrogative, but the speech function is typically either an offer or a command, though not a question (consider the response should a student answer *No*). Statements such as *I'm wondering if anyone has the answer,* intended as a question, can be misconstrued by students unfamiliar with this form of direction and control. 'Indirect' ways of construing the different speech functions differ greatly across cultures, even English-speaking cultures, and so use of indirect ways of regulating students has equity implications, as not every student will have the same understanding of the force of the teacher's statements, questions, commands and requests.

Analysis of *mood* and *speech function* can support analysis of the ways teachers direct students' attention and shape the development of a mathematical explanation. For example, Zolkower and Shreyar (2007) use analysis of *mood* and *speech function,* supplemented by comments on *modality*, to analyze the ways a teacher 'commands' her students in a U.S. urban sixth grade algebra classroom to 'think verbally' in a whole-class discussion as they notice patterns in a number array. Zolkower and Shreyar show how 'thinking' is constructed in language as the teacher organizes and scaffolds instruction. For the teacher, the key point of the lesson is how to conduct an open-ended search for patterns. They identify how the teacher mediates learning in powerful ways as she helps students recognize how to use thinking and talk about mathematics in the process of mathematical inquiry. Their analysis shows that students contribute more than half the clauses exchanged during the discussion, while at the same time the teacher exercises a lot of control, directing students to take notes on what is said and then use those notes as a model of how to solve such problems in the future. The teacher's linguistic choices enable each student's contribution to be considered, even when the 'right answer' has already been found. Addressing issues of authority, Zolkower and Shreyar argue that at the appropriate moments in a mathematics lesson, a directive, authoritative teacher can enable students' development of mathematical thinking.

This analysis illustrates the importance of studying language in context. As Zolkower and Shreyar point out, teacher commands are not always suited to such verbal mediation of mathematics and, at first glance, "the authoritative manner in which the teacher mediated the whole-class discussion may have seemed unconducive to engaging her students in genuine mathematics exchanges" (p. 200). In this instance, however, the analysis of the discussion shows that the commands guided students' participation. Teacher and students together engage in a conversation in which they are thinking aloud, with the teacher structuring the discussion to model an effective way to think about the algebra problem. Zolkower and Shreyar suggest that this analysis illustrates the power of a linguistic analysis "for studying the inner grammar of classroom interactions so as to illuminate the complexities and subtleties in the teacher's mediating role" (p. 200). Shreyar et al. (2010) offer a similar analysis with additional linguistic features in focus, in order to show how a sixth grade teacher in a Spanish-English bilingual classroom in Argentina supports an understanding that multiple approaches are possible as she engages students in a whole-class discussion of a problem about percentages.

O'Halloran (2004) demonstrates how analysis of mood and modality can shed light on interpersonal relationships in the classroom and on how students are positioned as learners in her analysis of three year 10 mathematics lessons in Perth, Australia: one taught to a class of male students at an elite private school; another taught to female students at an elite private school; a third taught to working-class students at a state school. She finds that the teacher and the male students in the private elite school interacted with each other much more directly than the teachers and students in the female and working-class schools, without the covert discipline strategies or use of sarcasm as a control mechanism that O'Halloran found in those latter contexts. She also reports that the male private-school students scored higher on national mathematics exams than the female private-school students, who in turn scored higher than the working-class students. She raises questions about how interaction in the mathematics classroom positions students in different ways, providing differential access to learning.

5 Discussion and Conclusions

Analysis of the discourse used in teaching and learning mathematics lets us look at the processes of teaching, learning and doing mathematics to understand the role that language choices play in pedagogical materials and classroom interaction. Several linguistic constructs have been described here, highlighting the ways they can help researchers illuminate equity issues. Analysis of *thematic patterns, process–participant configurations* and *modality* enables researchers to explore the integrity of mathematics itself and also how concepts are developed over time in pedagogical discourse, as well as to examine how the mathematics is presented; for example, whether as contingent or absolute. Analysis of *process–participant configurations, modality* and *mood/speech function* enables researchers to explore the processes through which knowledge is developed, focusing on the agency of students and the authoritativeness of the teacher, as well as the role of the teacher as mediator of learning, and how students are positioned as learners through classroom and pedagogical discourses.

In order to look at equity issues in grounded and comprehensive ways, it is important to recognize the role that language is playing in the particular context of use and to situate the analysis in a rich understanding of how a particular moment in a classroom relates to what has come before and follows. For example, it is not appropriate to say that a particular kind of modality always has a particular effect or that a teacher's moves should be of particular types in all instances. The discourse of teaching and learning is complex and evolving, with meaning emerging and shifting in ways that respond to the local context, always simultaneously construing the content and the interaction. And it is important to look at meaningful units of discourse in context to recognize how meanings of different kinds evolve as speakers interact and as writers develop ideas, since decontextualized examples tell us little about how the discourse is engaging and positioning participants or how the content is being presented and developed.

More research is needed to explore patterns of discourse in different settings and with diverse groups of students, analyzing the language used in mathematics classrooms and in mathematics pedagogical materials to illuminate the equity issues discussed in this chapter, as well as other issues of importance. We need more research that compares texts of different types, reveals problems with different wording and analyzes representations of mathematics at different levels and in different topics. We also need better understanding of how mathematics knowledge evolves over a unit of study of different topics at different levels to help teachers move between everyday and technical ways of making mathematical meaning. Analyses of the texts used for instruction and the ways teachers and students use and respond to those texts can identify challenges in the language and suggest ways to enable more effective presentation of content and how best to engage students in diverse contexts.

Through comprehensive investigation of how mathematical meanings are made, we can better identify the language structures that enable students to get access to the meaning being construed, the forms of interaction that are most effective in different types of classrooms and the language resources that are most relevant in constructing different mathematics topics at different levels and in different contexts. Close analysis of language is an important way to gain perspectives on processes of teaching, learning and doing mathematics. Using linguistic tools to draw insights from classrooms and contexts of different types, with a common language for exploring classroom discourse, can help us develop new understandings of the challenges students face, as well as develop for them new opportunities to learn and engage with mathematics in more equitable contexts.

A key issue in the study of equity is to identify forms of discourse that bring together the mathematics that enables students to achieve with ways of engaging with this mathematics that empower students to question critically – forms of discourse that afford students agency in their own learning and engage them with a mathematics that is open to exploration and discussion. This chapter has described linguistic tools that have been used to explore equity issues in mathematics, showing how close analysis of the actual language used in mathematics teaching and learning can reveal the meanings that are being construed.[4]

Related to the mathematics itself, researchers can explore the integrity of the mathematics presented and taken up in the classroom, the authoritativeness with which constructs are presented and the contingency with which definitions and other

[4]These linguistic tools can also be used for pedagogical purposes. Students can be made aware of how mathematics discourse works, engaging in critical discussion about how it positions them as learners. Teachers and students can engage in functional language analysis to unpack dense academic language, including the language of mathematics word problems and texts (Huang and Normandia 2008). Other linguistic tools are also available through SFL analysis – see González (2009) for an analysis of how conjunctions help a teacher structure an oral proof.

mathematics meanings are construed – whether as absolute or as open to questioning and challenges. Related to the learner, linguistic analysis can reveal differences in positioning of students and show how different contexts for learning mathematics may be more effective for particular groups of learners than others. The language analysis tools described here offer mathematics classroom researchers evidence that illuminates issues of *access, achievement, power* and *identity,* revealing the nature of the knowledge that teachers make available in mathematics classrooms and how students take up new knowledge in the context of actually doing mathematics, as well as the ways students are positioned and engaged as they learn.

Acknowledgement I would like to thank David Pimm for helpful comments in the development of this chapter, while not holding him responsible in any way for the final product.

Chapter 8
Mathematics in Multilingual Classrooms in South Africa: From Understanding the Problem to Exploring Possible Solutions

Mamokgethi Setati

In 1998, I started a research program focusing on mathematics education in multilingual classrooms in South Africa. The main problem that triggered its initiation had to do with my concern about the low mathematics performance of a majority of learners in multilingual classrooms in South Africa who learn in a language that is not their home language. At the core of this concern was a need to address the uneven distribution of mathematical knowledge and success.

Ten years have now elapsed and it is thus appropriate to reflect on the journey. During these 10 years, I worked with 15 mathematics teachers in Gauteng, North-West and Limpopo provinces. I video recorded, transcribed and analysed 60 primary and 70 secondary mathematics lessons in multilingual classrooms, lessons in which both teachers and learners are multilingual and none has the language of learning and teaching (LoLT) as their home language.

The purpose of this chapter is to describe my personal journey from understanding the problem to exploring a possible solution. In so doing, I highlight the contribution made to research and practice in the area of multilingualism and mathematics education. While it focuses on the exploration of one possible solution, I am under no illusion that there can only be one solution to the problem, given the range of multilingual contexts both in South Africa and elsewhere in the world. The chapter responds to the following questions:

- *What questions guided the journey and why?*
- *What is now known as a result of the journey?*
- *What is still not known?*

M. Setati (✉)
College of Science, Engineering and Technology, University of South Africa, Pretoria, South Africa
e-mail: setatrm@unisa.ac.za

B. Herbel-Eisenmann et al. (eds.), *Equity in Discourse for Mathematics Education: Theories, Practices, and Policies*, Mathematics Education Library 55,
DOI 10.1007/978-94-007-2813-4_8, © Springer Science+Business Media B.V. 2012

1 Understanding the Problem

Although the concern about the mathematics performance of learners in multilingual classrooms in South Africa served as inspiration for the research, investigating the impact of multilingualism on learner performance is not a straightforward matter. This is mainly because learner performance and, by implication, mathematical achievement is not just determined by language proficiency, but also by a complex set of interrelated factors. So while proficiency in the language of learning and teaching is one of the factors that impacts learner performance, it is not the only factor. I therefore decided to develop an understanding of the problem of poor learner performance in multilingual mathematics classrooms by exploring the following questions:

- *How does the fact that learners learn mathematics in a language that is not their home language shape their mathematics learning?*
- *What is it that teachers in multilingual classrooms do to support their learners' understanding of mathematics?*

It is widely accepted that language is important for learning and thinking and that the ability to communicate mathematically is central to learning and teaching school mathematics (see, for example, Adler 2001; Moschkovich 1996, 1999, 2002; Pimm 1987; Sfard et al. 1998; Setati 2005a, b). Part of learning mathematics is acquiring fluency in the language of mathematics, which includes words, phrases, symbols and abbreviations, and fluency in a range of discourses that are specific to mathematics. The relationship between language and mathematics learning, however, takes on a specific significance in multilingual classrooms in which learners learn mathematics in a language they are not fluent in. Mathematics teachers in these classrooms have a dual task of continuously needing both to teach mathematics and to develop the learners' fluency in the LoLT (e.g. English) at the same time. Learners, on the other hand, have to cope with the new language of mathematics as well as the new language in which mathematics is taught (English). Thus, learning and teaching mathematics in multilingual classrooms involves managing the interaction between formal and informal mathematics language; between procedural and conceptual discourses; between ordinary language and mathematical language; between the home language and the LoLT.

Formal mathematics language is about acting–interacting–thinking–speaking–reading–writing in mathematically appropriate ways, which will enable the learners to understand and be understood by other members of the wider mathematics community (Gee 1999). Informal language is the kind that learners use in their everyday interactions, sometimes to communicate mathematical ideas encountered in everyday life. Both formal and informal mathematics languages are carried by distinctive mathematics discourses: of relevance to this chapter are procedural and conceptual discourses. Procedural discourse involves interactions that focus on the procedural steps taken to solve a problem. This discourse is produced by a computational orientation to teaching, where mathematics is viewed as being composed of procedural steps and doing mathematics as computing or following a set of procedures in the absence of any reason for the computation. Conceptual discourse refers to interactions in which the reasons for calculating in particular ways and using particular procedures to solve a mathematical problem become explicit topics of conversation (Sfard et al. 1998, p. 46).

Doing mathematics involves an ability to engage in a range of discourses. Furthermore, mathematics teaching and learning occurs in a mixture of ordinary language and mathematical language. According to Pimm (1987, p. 88), the learners' failure to distinguish between the two can result in breakdowns in communication. In a multilingual classroom of English language learners, the confusion between ordinary English and mathematical English is complicated by the fact that both languages are new to the learners. Consider, for instance, the excerpt below from a publication by French mathematician, Ahmed Laghribi:

> *Proposition 5.1 Soit A une F-algèbre simple centrale de degré* 2 *munie d'une involution orthogonale* σ *anisotrope. Soit* φ *une forme quadratique sur F anisotrope de dimension* ≥ 2. *Alors,* σ *devient isotrope sur* $F(\varphi)$ *si et seulement si* φ *est semblable à* $[1] \perp [d]$ *où disc* $\sigma = dF^{*2} \in F^{*}/F^{*2}$. (2005, p. 170)

This extract above would not be easy to understand for non-French speaking mathematicians. In the same way, the excerpt below from an English mathematics textbook would not make much sense to an English-speaking non-mathematician:

> Let A be a σ-algebra of subsets of X and μ, ν two finite measures on A. Then ν may be expressed uniquely as $\nu = \nu_1 + \nu_2$ where $\nu_1 \ll \mu$ and $\nu_2 \perp \mu$. (Weir 1974, p. 219)

While this excerpt is written in English, it is in the mathematics register and not ordinary language. Thus, any teacher of mathematics should consider the fact that while the mathematics language can be in English, it is not necessarily easy for English speakers who are not mathematicians to understand it. In a multilingual classroom, it is important to consider that in addition to learning mathematics, which has its own register, learners are also learning English and thus experience the double challenge of learning the mathematics register in a language that they are still learning.

2 Analysing Empirical Data to Understand the Problem

Analysis of lesson observation data collected from 1998 to 2003 with the purpose of understanding the problem showed a dominance of English and the use of procedural discourse. The dominance of English was evident in the interactions between learners as well as with the teacher. The fact that the interactions were in English is not only the only thing that characterised the interactions, but also the fact that they were abbreviated and procedural (for a detailed discussion, see Setati 2005a).

Procedural discourse was evident in the mathematics tasks and tests that the teachers gave learners, as well as in the interactions that occurred during the observed lessons, which suggested that conceptual discourse was not valued as mathematical knowledge. Below are examples of two tasks given to learners in some of the classrooms observed.

> In the SPCA are twelve cages; in each cage are twelve dogs. How many dogs are there altogether? (A task given to grade four learners in 1998.)

> Use algebra to solve the following equation, $x+2=8$. (A task given to grade eleven learners in 2003.)

Although the grade four task is about multiplication of two-digit numbers, it is presented as a word problem. It remains, however, a low-level task in terms of cognitive demand, since it can easily be done by learners reproducing a previously learned procedure for multiplying two-digit numbers. On the other hand, although the grade 11 task stipulates the use of algebra, it is obvious that learners would not need any algebraic procedure to solve it, since there is no ambiguity in the task and it can be easily solved by inspection. These tasks lack connection to the mathematical concepts or meaning that underlie the procedure to be followed: instead of developing mathematical understanding, their focus is on producing correct answers (Stein et al. 2000). Mathematics tasks and tests given to learners inevitably communicate to learners what is valuable mathematical knowledge. I therefore regarded the absence of high cognitive demand tasks that demand fluency in conceptual discourse in the lessons observed as problematic.

The learners' lack of exposure to high-level cognitive demand mathematical tasks, as well as the dominance of English despite learners' limited fluency in it, and accompanied by the prevalence of procedural discourse, raised the following question: *What shapes the nature of the mathematics tasks and the language choices made in these classrooms?* At the heart of this question was a seeming disjuncture between what research and policy recommends and the practices that I was observing in the classrooms.

On the one hand, the South African Language in Education Policy (LiEP) recognises 11 official languages and encourages multilingualism, as well as language practices such as code-switching, as resources for learning and teaching in multilingual classrooms. Research argues that classroom interactions that include the use of the learners' home languages can support mathematics learning (Adler 2001; Khisty 1995; Moschkovich 2002). On the other hand, there was a dominance of the use of English for teaching and learning in these classrooms, which seemed to be inconsistent with the recommendations from policy and research. To understand this seeming disjuncture, it was important to explore, through individual interviews with teachers and learners, the language(s) they prefer for teaching and learning mathematics and consider how they shape the mathematics tasks selected and interactions in these classrooms.

During the interviews teachers were asked, "Which language do you prefer to teach mathematics in? Why?" Through the analysis of data, over and above all else, *English is international* emerged as a dominant discourse that shaped the teachers' language choices. All six teachers interviewed stated both ideological and pragmatic reasons for their preference for teaching mathematics in English.

These teachers were aware of the linguistic capital of English and the symbolic power it bestows on those who can communicate in it. One of the teachers, for example, said, "I prefer to teach in English because it is a universal language". All of the teachers used similar wording, referring to English as an 'international' or 'universal' language. Awarding such a status to English suggests that they see English as being 'bigger than' themselves. They do not have any control over the international nature of English. All they can do is to prepare their learners for participation in the international world, and teaching mathematics in English is an important part of this preparation. One of the teachers expressed the reasons for her preference for English as follows: "it is an international language [...] The textbooks are written in English, the question papers are in English ..." Another one argued that, "If they do not learn the language, how will they be able to cope in higher classes?" None of the reasons that the teachers gave for their preference for English were related to mathematics learning; instead, they were about the need to ensure that learners can gain access to the social goods that fluency in English makes available.

Analysis of interview data highlighted the teachers' preference for English as the language of learning and teaching mathematics (Setati 2008). A glaring absence was any reference to how learning and teaching in English, which they prefer, would promote their learners' access to mathematics knowledge and success. The teachers regarded teaching mathematics in English in these multilingual classrooms as another opportunity for learners to gain access to English. Explanations for preferred language(s) for mathematics teaching focused on English and not mathematics. These teachers positioned themselves in relation to English (and so socioeconomic access) and not mathematics (i.e. epistemological access). For a detailed discussion of this positioning, see Setati (2008).

All of the interviewed learners are multilingual, since they have fluency in at least four languages. At the time of the study, they were all in grade 11 and studying mathematics in English, which is not their home language. Learners were given an opportunity to choose their preferred language for the interview. All of the learners chose to be interviewed either in their home language or in a mixture of English and their home language. All the questions asked during the interview were therefore in their language of choice. The main question they were asked was: *Which language or languages do you prefer to be taught mathematics in? Why?*

Although there were conflicting discourses in the learners' views, what was clear was that the majority of learners expressed their preference to be taught mathematics in English. For these learners, learning mathematics in English is not so much about choice, it is just how things should be. One learner said, "English is an international language; just imagine a class doing maths with Setswana", while another said, "it is the way it is supposed to be, because English is the standardized and international language". For these learners it is unimaginable for mathematics to be taught in any other language. The use of English as a language of learning and teaching mathematics is simple commonsense to them; they cannot imagine mathematics without English.

Among their reasons why they want to be taught in English is the fact that mathematics textbooks and examinations are in English, university lecturers and job interviews are only in English and communication with 'white people' is in English. All these factors contributed to the discourse that without fluency in English a learner would not have access to significant social goods such as higher education and employment. Like teachers, these learners saw mathematics learning as another opportunity for gaining fluency in English. This was the case even for the two learners who indicated that for them it does not really matter which language is used for teaching and learning, because mathematics is a language in its own right. Below is an extract from the interview with one of the two learners.

Researcher: So if you had a group of learners who want to do maths in IsiZulu, what would you say to them?

Lehlohonolo: That's their own problem, because if they get out of high school, they cannot expect to find an Indian lecturer teaching maths in IsiZulu. English is the simplest language that everyone can speak, so they will have to get used to English whilst they are still here.

Throughout the interview, Lehlohonolo never connected success in mathematics or lack of it to fluency in English; however, in the above extract he argues for the importance of gaining fluency in English before completing high school. The sentiment that English is bigger than us – and thus cannot be avoided or ignored because in higher education no lecturer will be able to teach mathematics in an African language – is evident in Lehlohonolo's discourse.

Despite the overwhelming discourse that foregrounds the hegemony of English and the need to gain access to social goods that English makes possible, there are nevertheless differences in the manner in which different learners positioned themselves. The learners who explicitly indicated that it does not really matter what language mathematics is taught in positioned themselves in relation to mathematics. Their language preferences were connected to gaining proficiency in mathematics rather than gaining fluency in English. The rest of the learners positioned themselves in relation to English in the sense that they were more concerned with gaining fluency in English, so that they can access employment and higher education. Their desire to gain fluency in English was not connected to improving their mathematics learning, but to the possibility of accessing social goods. As a result, they saw mathematics teaching and learning in multilingual classrooms primarily as an opportunity to gain fluency in English.

3 What Does This Mean for Research, Policy and Practice?

Given the hegemony of English both in South Africa and elsewhere in the world, the teachers' and learners' preference for English is not surprising. There is no doubt that English is a valued linguistic resource in South Africa. While the new language policy in South Africa is intended to address the overvaluing of English and Afrikaans[1] and the corresponding undervaluing of African languages, in practice English continues to dominate. Although it is the main language of a minority, English is both the language of power and the language of educational and socio-economic advancement; that is, in South Africa, it is a dominant symbolic resource in the linguistic market (Bourdieu 1991).

The linguistic market is embodied by and enacted in the many key situations (e.g. educational settings, job situations) in which symbolic resources, like certain types of linguistic skills, are demanded of social actors if they want to gain access to valuable social, educational and eventually material resources. The symbolic power of English has given rise to a sense that the purpose of school is to teach English. This identification of schooling with the learning of an additional language (English) enabling wider communication is not unique to South Africa. Benton (1978, p. 126), for example, describes the same attitude to formal education among the Maori in New Zealand. The same situation exists in Peru (Hornberger 1988), where education has always meant learning Spanish (a second language enabling wider communication).

The identification of English with schooling in South Africa is therefore not surprising, since English is ideally suited to both functions of education there: it is both the vehicle of acculturation and an easily identifiable trait for maintaining privilege. In the same way, limited fluency in English is an easily identifiable trait of lower status and disadvantage. The dominance of English in school – and in multilingual mathematics classrooms in particular – is a reflection of the status that has been given to this language outside school.

The above discussion indicates that decisions about which language to use in multilingual mathematics classrooms, how and for what purposes, are not only pedagogic but also political (Setati 2005b). This conclusion explains why recommendations from policy and research are hard to translate into practice in multilingual contexts such as South Africa. An assumption embedded in the South African Language in Education Policy (LiEP) is that multilingual mathematics teachers, learners and parents are somehow free of economic, political and ideological constraints and pressures when they apparently freely opt for English as the LoLT. The South African LiEP seems to be taking a structuralist and positivist view of language, one that suggests that all languages can be free of cultural and political influences.

Research, on the other hand, recommends the use of the learners' home languages and thus suggests that concerns about learners' access to mathematics precede concerns about access to social goods, such as tertiary education and jobs. However, evidence from the data shows that calls by students, teachers and parents for social access predominate over those for cognitive access. The challenge here is that most of the research in this area of study is framed by a conception of mediated

[1] Afrikaans is one of South Africa's 11 official languages. It developed out of Dutch settlement. During the apartheid era, English and Afrikaans were the only two official languages.

learning, where language is seen as a tool for thinking and communicating. Although it appropriately foregrounds the mathematics, it does not consider the political role of language. There is, thus, a need for research in this area of study to recognise and acknowledge language as political, because without such recognition we will fail to understand and work with the demands that teachers and learners in multilingual classrooms experience.

As explained earlier, lesson observation data in this study highlighted the dominance of English and procedural discourse in these multilingual classrooms. Other researchers have interpreted this dominance of procedural teaching as a function of the teachers' lack or limited knowledge of mathematics (Taylor and Vinjevold 1999). What the above analysis and discussion suggests is that the problem is much more complex. There seems to be a tension between the desire to gain access to English and the important, but not always recognised and acknowledged, need to gain access to mathematical knowledge. It therefore seems that it is not necessarily true that teachers' mathematical knowledge is limited or lacking, but rather that access to conceptual explanations in English is limited or lacking.

Many of these teachers have themselves learnt mathematics at school and at university or college via a procedural mathematics discourse in English and have had little exposure to conceptual mathematical discourse, either in their home language or in English. While language and knowledge are intricately connected, limited fluency in mathematical conceptual discourse does not necessarily equate to lack of mathematical knowledge, but is, of course, a stumbling block in the use of that knowledge nevertheless. It is not surprising that, in their teaching, teachers seem to be experiencing a tension between the desire to make English accessible to their learners and the important but not always recognised and acknowledged need to ensure that the learners also gain access to mathematical knowledge.

The analysis and discussion above suggests that a possible solution to the problem that guided the research programme that started in 1998 can be found in an exploration of ways of drawing on the learners' home languages while ensuring that they gain fluency in English. This calls for considering the learners' home languages as a resource for teaching and learning. The challenge, however, is that in a context such as South Africa, where the hegemony of English is so prevalent, regarding learners' home languages as a resource tends to be seen as a threat to multilingual learners' development of fluency in English.

As Sachs (1994) has pointed out, in South Africa "all language rights are rights against English" (p. 1). Hence my argument that, for the use of the learners' home languages in the teaching and learning of mathematics in multilingual classrooms to be successful, it must ensure that learners gain epistemological access without losing access to English (Setati et al. 2008). Granville et al. (1998) present a similar idea in relation to the South African language in education policy, where they argue for English without g(u)ilt. What is new about the proposed use of language(s) in exploring one possible solution presented in this chapter is the different orientation it brings, by focusing on learning and teaching rather than policy.

4 Theoretical Underpinnings for One Possible Solution

Both in research and in the public domain, debates on language and mathematics teaching and learning in multilingual classrooms tend to create dichotomies of language choices and theoretical perspectives (see, for example, Setati 2005b). These dichotomies create an impression that using the learners' home languages for teaching and learning must necessarily exclude or be in opposition to English and developing the learners' mathematical proficiency must necessarily be in opposition to developing fluency in English. Furthermore, these dichotomies create an impression that doing research that is informed by a sociopolitical perspective should necessarily exclude or be in opposition to cognitively oriented research. The argument presented below shows that this is indeed a false dichotomy and that a possible solution to the problem can start by acknowledging the complex relationship between language choices, looking for synergy between the language of power (English) and other languages crucial for supporting conceptual understanding. Rejecting these dichotomies is the starting point and the data that I present in this chapter makes a start at exploring this synergy as a possible solution to the problem.

In a Sunday newspaper article entitled 'Why don't kids learn maths and science successfully?', reprinted in the *Science in Africa* magazine (SIA 2003), Sarah Howie, a South African researcher, is quoted as saying that the most significant factor in learning mathematics is not whether the learners are rich or poor. It is whether they are fluent in English. She insisted:

> Let's stop sitting on the fence and make a hard decision. We must either shore up the mother tongue teaching of maths and sciences, or switch completely to English if we want to succeed. (p. 1)

She made this argument drawing on her analysis of South Africa's poor performance in the Third International Mathematics and Science Study of 1995 (see also Howie 2003, 2004). While Howie's downplaying of the effects of socioeconomic class on mathematics education and achievement is problematic, of relevance here are her views about language choice and use in multilingual mathematics classrooms. In the light of the discussions above, Howie's suggested solution to the problem is simplistic. Multilingual learners ought to be viewed in a holistic manner, which is different from Howie's monolingual view. Multilingual learners have a unique and specific language configuration and therefore they should not be considered as the sum of two or more complete or incomplete monolinguals. The possible solution explored in this chapter is informed by this holistic view of multilingual learners.

One of the lessons that emerged from the process of understanding the problem in my research is the fact that it is not productive to separate cognitive matters from the sociopolitical issues relating to language and power when exploring the use of language(s) for teaching and learning mathematics in multilingual classrooms. Although cognitively oriented research does not deal with the political role of language and sociopolitical issues relating to the context in which teaching and learning takes place, it is important to acknowledge that it does attend to issues relating to the quality of the mathematics and its teaching and learning in multilingual classrooms. It is thus important to work against the dichotomies, not only of language choices but also of theoretical perspectives.

The exploration of a possible solution was broadly informed by an understanding of language as, in Lave and Wenger's (1991) terms, 'a transparent resource'. Although this notion of transparency is not usually applied either to language as a resource or to learning in school, it is illuminating of language use in multilingual classrooms (see also Adler 2001). Lave and Wenger argue that access to a practice relates to the dual visibility and invisibility of its resources:

> Invisibility is in the form of unproblematic interpretation and integration into activity, and visibility is in the form of extended access to information. This is not a dichotomous distinction, since these two crucial characteristics are in a complex interplay, their relation being one of both conflict and synergy. (p. 103)

For language to be useful in any classroom, it must be both visible and invisible: visible so that it is clearly seen and understood by all; invisible in that when interacting with written texts and discussing mathematics, this use of language should not distract attention from the mathematical task under discussion, but instead facilitate the learner's mathematics learning. This idea is similar to the use of technology in mathematics learning. The technology needs to be visible, so that the learners can notice and use it. However, it also needs to be simultaneously invisible, so that the learners' attention is focused on the mathematics problem that they are trying to solve. Like technology, language needs to be a transparent resource. As Lave and Wenger argue, the idea of the visibility and invisibility of a resource is not a dichotomous distinction: it is not about whether to focus on language or mathematics, it is about recognising that the two are intertwined and are constantly in complex interplay. This constant interplay is even more intricate when teaching and learning mathematics in a multilingual classroom of learners who are still learning the LoLT.

Good teaching is critical to learning and improving learner performance. It is thus important to explore relevant teaching strategies for multilingual mathematics classrooms that take into consideration the constant complex interplay between language and mathematics. Lave and Wenger's concept of transparency is useful in conceptualising language use in teaching and learning mathematics in multilingual classrooms, which are characterised by complex multiple teaching demands: the learners' limited proficiency in the language of learning and teaching (English); the challenge to develop the learners' mathematical proficiency; the presence of multiple languages.

Recent research that explores productive pedagogies (Hayes et al. 2006) and ways in which teachers can develop learners' mathematical proficiency (Kilpatrick et al. 2001) through use of mathematical tasks with different cognitive demands (Stein et al. 2000) does not consider language use and thus ignores the fact that some of the learners are not fluent in the language in which they learn mathematics. The possible solution explored in this chapter recognises the complexity of teaching and learning mathematics in multilingual classrooms, where the challenges of developing learners' mathematical proficiency are intertwined with challenges of the learners' proficiency in the LoLT (English). This solution is guided by two main principles, which are informed by the theoretical assumptions elaborated in the discussion above.

1. It involves the *deliberate* use of the learners' home languages. The word 'deliberate' is emphasised, because with this strategy the use of the learners' home languages is deliberate, proactive and strategic, and not spontaneous and reactive as happens with code-switching.
2. It involves the selection of interesting and challenging real-world mathematical tasks, through which learners would develop a different orientation towards mathematics than they have had and would be more motivated to study and use it (Gutstein 2003). Many learners in multilingual classrooms in South Africa have what he describes as "the typical and well-documented disposition with which most mathematics teachers are familiar – mathematics as a rote-learned, decontextualised series of rules and procedures to memorize, regurgitate and not understand" (p. 46). In this exploration, high-cognitive-demand tasks were used, tasks that present real-world problems which the learners can find interesting and with which they can usefully engage (Stein et al. 2000).

5 The Exploration

Piloting was done in five grade 11 multilingual mathematics classrooms. This chapter focuses on data collected in Terence's classroom in a multilingual high school in Soweto, Johannesburg. There were 36 learners in his class who had the following home languages: Setswana, Xitsonga, IsiZulu and Tshivenda. Each of the learners was able to communicate in at least four languages and they were learning English as an additional language, as well as their respective home languages as subjects. Terence is multilingual and fluent in eight languages,[2] including all the home languages of his learners as well as English. His home language is Setswana. At the time of the study Terence had been teaching mathematics at secondary school level for 15 years.

Data was collected through lesson observations and individual learner interviews. Lessons were observed and video-recorded for four consecutive days. At the end of the four days, four learners from different home language groups were interviewed by another researcher who had not been present during the lesson observations. The interview focused on the learners' reflections and views about the lessons.

[2] This kind of multilingualism is not unusual in Johannesburg, South Africa. Given the integration of different ethnic groups, a majority of black South African teachers (indeed black South Africans in general) in the Gauteng province are multilingual and can communicate in at least four languages.

Below is a task that learners were working on during the lessons observed and analysed in this chapter. This task was translated into the four home languages of the learners in the class.

Cost of electricity (selected from Malati Materials, MALATI 2005)

The Brahm Park electricity department charges R40,00 monthly service fees then an additional 20c per kilowatt-hour (kWh). A kilowatt-hour is the amount of electricity used in one hour at a constant power of one kilowatt.

1. The estimated monthly electricity consumption of a family home is 560 kWh. Predict what the monthly account would be for electricity.
2. Three people live in a townhouse. Their monthly electricity account is approximately R180,00. How many kilowatt-hours per month do they usually use?
3. In winter the average electricity consumption increases by 20%, what would the monthly bills be for the family home in (1) above and for the townhouse?
4. In your opinion, what may be the reason for the increase in the average electricity consumption in (3) above?
5. Determine a formula to assist the electricity department to calculate the monthly electricity bill for any household. State clearly what your variables represent and the units used.
6. (a) Complete the following table showing the cost of electricity in Rand for differing amounts of electricity used:

Consumption (kWh)	0	100	200	300	400	500	600	700	800	900
Cost (in Rand)										

   (b) Draw a graph on the set of axes below to illustrate the cost of different units of electricity at the rate charged by the Brahm Park electricity department.

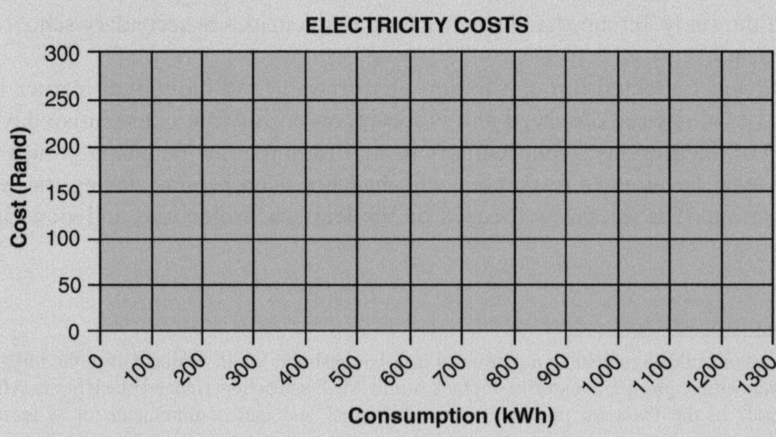

(continued)

Cost of electricity **(continued)**

After careful consideration, the electricity department decided to alter their costing structure. They decide that there will no longer be a monthly service fee of R40,00 but now each kilowatt-hour will cost 25c.

7. What would be the new monthly electricity accounts for the family home and the townhouse?
8. (a) Complete the following table showing the cost of electricity in Rand for differing amounts of electricity used using the new costing structure:

Consumption (kWh)	0	100	200	300	400	500	600	700	800	900
Cost (in Rand)										

   (b) Draw a graph on the same set of axes in question 6(b) to illustrate the cost of electricity for different units of electricity using the new costing structure.
9. Do both the family home and the townhouse benefit from this new costing structure? Explain.
10. If people using the electricity had the option of choosing either of the two costing structures, which would you recommend? Clearly explain your answer using tables you have completed and graphs drawn in questions (6a) and (6b) and (8a) and (8b) above.

During the lessons, learners were organised into seven home language groups: two Setswana groups, two Tshivenda groups, two IsiZulu groups and one Xitsonga group. Six of the groups had five learners and one group had six learners and they were given tasks in two language versions (English and their respective home language). Learners were explicitly made aware of the two language versions of the task and encouraged to communicate in any language including their home languages at any stage during the lessons.

All the lessons and learner interviews were video-recorded and then transcribed. Presences (what was visible) and absences (what was invisible) in what the learners were talking about were focused on during analysis of the video-recording and the transcribed data. In the lesson observation data, what was most visible were the learners' attempts to find possible solutions to the questions in the task without much focus on the language. There was only one incident during the lessons observed where language became visible but not simultaneously invisible in one of the groups. Language was, however, constantly visible for Terence, the teacher: for instance, when asking learners to read he would specify in which language they should read.

5.1 When Language Was Visible and Invisible

The analysis shows that when language was transparent, learners' interactions were conceptual – that is, learners' interactions were focused not only on what the solution was but also why it was correct. The two extracts below took place in the Tshivenda group and are typical of how language functioned as a transparent resource during interactions between Terence and the learners and also between the learners themselves. Both extracts were taken from the first lesson at the time when the learners were beginning their work on the task and they needed to understand the following statement in the problem:

> The Brahm Park electricity department charges R40,00 monthly service fees then an additional 20c per kilowatt-hour (kWh). A kilowatt-hour is the amount of electricity used in one hour at a constant power of one kilowatt.

The extract below shows the interaction between Terence and the learners in the group. Here Terence is working with them on the two charges mentioned in the problem, the R40 monthly service fee and the additional 20c per kWh.

1	Terence:	Forty rhanda heyi, vhoibadhala when **[When is the forty rand paid?]**
2	Sipho:	In a month.
3	Terence:	Twenty cents yone **[What about the twenty cents?]**
4	Given:	Twenty cents yo ediwa. **[Twenty cents is added].**
5	Terence:	Why i ediwa **[Why is it added?]**
6	Learners:	*(Silent).*
7	Terence:	Vhoi edela mini? Twenty cents vhoi edela mini **[Why is it added? Why is twenty cents added?]**
8	Learners:	*(Inaudible).*
9	Terence:	Okay, if you use electricity ukho bhadala forty rand? **[Okay, if you use electricity will you pay forty rand?]**
10	Learners:	Yes meneer **[sir]**.
11	Terence:	If unga shumisanga electricity ukho bhadala forty rand **[If you did not consume electricity, will you pay forty rand?]**
12	Sipho:	No, no no …
13	Given:	Haena, whether ushumisile ore haushumisanga, ukhobhadala forty rhanda **[No, whether you have consumed electricity or not, you pay the forty rand].**
14	Terence:	Whether ushumisile ore haushumisanga? **[Whether consumed or not?].**
15	Sipho:	Eya, yes, it is a must.
16	Terence:	It is a must?

One very noticeable thing about the extract above is the fact that it is in a mixture of English and the learners' home language (Tshivenda). This, as indicated earlier, was typical of interactions during the lessons observed in this class. The unproblematic move between Tshivenda and English without explicit negotiation between interactants is an indication that language is functioning as a transparent resource. Whilst language is visible, in the sense that learners recognise which language is used, it is also invisible enough to be used without distracting attention from the task. This invisibility of language as a mediating tool allows focus on and thus supports visibility of the mathematics the learners are discussing (Lave and Wenger 1991). At the same time, the visibility of language (i.e. tasks given in two languages) is necessary for allowing its subsequent, unproblematic, invisible use.

In the extract above, learners were struggling to understand the phrase 'an additional 20c per kilowatt-hour (kWh)'. Although they understood that everyone who has electricity is supposed to pay the R40 monthly cost, and also that 20c is added, they seemed to be having difficulty in understanding why the 20c is added. In turns 5–7 Terence asked them why 20c was added. Seeing that they were not able to answer the question, he moved back to asking them about the R40 in turn 9. By doing this, Terence was separating the R40 from the 20c, so that the learners could see that, while everyone who has electricity is required to pay R40, how much they pay thereafter depends on the number of kilowatt hours they used and 1 kWh costs 20c. The extract above ends with Terence in turn 16 having established with the learners that the R40 payment is a mandatory service fee for everyone who has electricity. In the extract below, the learners carry on with the discussion (on their own) about when and why 20c is added.

1 Given:	Hei, nayo ... ar ... *(Giggles)* ... So forty rhanda hi monthly cost ne, then ba yieda nga twenty cents kha kilowatt for one hour. Then after that, angado shumisa ..., baibidza mini? Heyi ... ndoshumisa one kilowatt nga twenty cents kha one hour **[Hey, this question ... ar ... *(giggles)* ... So forty rand is the monthly cost, then they add twenty cents per kilowatt-hour. ..., they use..., what do they call it? Hey ... they use one kilowatt-hour for twenty cents]**.
2 Sipho:	Eya **[Yes]**.
3 Given:	Boyieda, maybe boshumisa twenty cents nga one hour **[They add it, maybe they use twenty cents per hour]**.
4 Sipho:	Eya, yantha **[Yes, one hour]**.
5 Given:	Iba ... **[It becomes...]**.
6 Given and Sipho *(together)*:	Forty rand twenty cents
7 Sipho:	Yes, vhoibadela monthly, ngangwedzi ya hona. Yo fhelela, yes. Sesiyaqubheka. **[Yes, they pay it monthly, each month. It is complete, yes. We continue.]**

The transparent use of language continues in the above extract. The learners' focus was not on which language(s) they were using; they were focused on communicating their understanding. This transparency of language enabled conceptual interactions between the learners and the teacher and also among learners themselves. Using their home language, Tshivenda, as a legitimate language of interaction together with English made it possible for them to understand that in this case 1 kWh costs 20c. The learners were not concerned with the correctness of their grammar in Tshivenda or English, they were more focused on gaining an understanding of the problem and having both language versions served as resources that they could draw on as and when they needed to.

5.2 *Language as a Transparent Resource in the Learners' Reflections on the Lessons*

As indicated earlier, reflective individual interviews were conducted with learners after the first week of lessons. Terence selected four learners from different home language groups for interviewing, so as to get their reflections on the lessons observed. The interviews were semi-structured and were conducted by a research assistant who used to be a teacher and the head of the mathematics department in the school. This assistant was not part of the team that collected the lesson observation data. The assumption was that this familiarity would help learners to talk frankly about their experiences of the lessons. Learners were given an opportunity to select languages they wanted to be interviewed in. The interviews were video-recorded and then transcribed.

The analysis of the learner interviews focused on what was visible, i.e. presences (what the learners were talking about) and what was invisible, i.e. absences (what they were not talking about). The expectation was that the two main changes introduced (language and the nature of the task) would be most visible in what the learners would talk about during the interviews. In the extracts below from interviews with individual learners, the interviewer asked them the same open question about what was happening in their class:

Interviewer:	I understand this week you had visitors in your class. What was happening?
Sindiswa:	Er…, we were learning a lesson in which we can calculate electricity er …. amount … er … the way in which the electricity department can calculate the amount of electricity unit per household.
Nhlanhla:	We were learning about how to calculate …er…er… kilowatts of the electricity, how do we … like … how can we calculate them and when … at …, Besifunda mem ukuthi ugesi udleka kakhulu nini. **[We were learning about when there is high electricity consumption.]**
Colbert:	Er …we were just solving for electricity, kilowatt per hour, for comparing if they are using card or the meter, which is both, I think are the same.
Sipho:	Er, the visitors they were doing research. Gošho gore ba sheba gore bothata … bothata ba rona bo mo kae, ka … ka … maths, then they found out that er… ba bang ha ba understende dilanguage, like English, so then ha ba botsa karabo then they can't find the answer. **[They were checking what our problems were, with … with … maths, then they found that er … some of us do not understand the languages, like English, so then when they ask a question then they can't find an answer.]** So Mr Molefe then decided to … to … make it in … in English and vernacular language to … to …, for us to understand.

Three of the learners above pointed to the mathematical task that they were working on during the lessons, thus suggesting that the task is what they found as central to what was happening during the lesson. As explained earlier the approach explored in this class centres around two principles: (1) the deliberate use of the learners' home languages; (2) the selection of interesting and challenging real-world mathematics tasks. Given these principles, it was expected that the use of the learners' home languages in the tasks given would be the most prominent thing for the learners to notice. What is emerging in the extracts above is that for the learners the context of the task, cost of electricity, was more prominent.

Given the expectation that the learners would point to language as most prominent about the lessons, the interviewer probed further as below:

Interviewer: But what was so special about the lessons?

Sindiswa: It does not include those maths … maths. It is not different, but those words used in Maths didn't occur, didn't occur but we weren't using them. … Er … 'simplifying', 'finding the formulas', 'similarities', …

Nhlanhla: Hayi, no mem, ku-different… Okokuqala mem, ilokhuza, la sidila ngama-calculations awemali, manje ku-maths asisebenzi ngemali.
[No mam, it is different. Firstly mam, we were working with money and usually in maths we do not work with money.]

Colbert: Iya, basenzele in order to … ukuthi ibe simple and easy to us, because most of people, uyabona, aba-understendi like i … like i-card ne meter. Abanye bathi i-meter is … i-price yakhona i-much uyabona, i-card iless i-price yakhona, that's why uyabona. So, abantu abana-knowledge, uyabona, bakhuluma just for the sake of it. So, I think for us, because we have learnt something, both are the same.
[Yes, you see they made it easy for us, because most people do not understand, like card or using a meter. Some say when using the card you pay less than when using the meter, you see. So people do not have knowledge out there, they just talk for the sake of it. So think, for us we have learnt something, both are the same.]

Sipho: Gošho gore ba sheba gore bothata … bothata ba rona bo mo kae, ka … ka … maths, then they found out that er… ba bang ha ba understende dilanguage, like English so, then ha ba botsa karabo then they can't find the answer. So Mr Molefe then decided to … to … make it in … in English and vernacular language to … to …, for us to understand.
[They were looking at the problem… where our problem is, with… with… maths, then they found that er … some of us do not understand languages like English, so when they ask for the answer we can't find it. So Mr. Molefe then decided to… to … make it in English and vernacular language to … to… for us to understand.]

In responding to the interviewer's question above, both Sindiswa and Sipho point to language. Sindiswa points to how the task differs from the textbook tasks that they are used to. Sindiswa says of the observed lesson that the absence of many of the terms usually associated with the mathematics classroom was significant, even though the essence of the lesson and activity remained unchanged. It is evident that for Sindiswa language played a clear role in the 'feel' of the lesson. This is echoed by Sipho, whose response and choice of language is very interesting. What stood out for him about the lessons was what the introduction of the learners' home languages allowed them to do. It changed the dimensions of the interaction, increased participation and intervened at the level of meaning. Noticeably, he does *not* say, "ha ba botsa karabo then <u>they don't know</u> the answer" ["when they ask for the answer <u>they don't know</u> the answer"], he says, "ha ba botsa karabo then <u>they can't find</u> the answer" ["when they ask for the answer <u>they</u>

can't find the answer"]. In Sipho's analysis, the learners may *have* the answer in one language, but their inability to *find* it in another language (English) has direct effects on their participation and performance. Thus, mathematical knowledge and inability to articulate it in English are not the same thing, even while they can be complexly intertwined.

On the face of it, Sindiswa and Sipho address different aspects of the changed lesson. However, both highlight the manner in which the use of language (or the absence of certain kinds of language) can either enable comprehension or constrain learning. Both see the actual mathematical activity as unchanged. For Sindiswa, when 'difficult' words are minimised, then learners and teachers can get on with the usual business of mathematics, focusing on the task and allowing learners to experience mathematics differently and more fluently.

Nhlanhla and Colbert point to the nature of the task. For Nhlanhla, what stood out the most is the fact that in mathematics they usually do not deal with calculations involving money and so these lessons were special because they involved money calculations. This resonates with Colbert's focus on the value of the task beyond the lesson. For him, it was about clarifying a real-life situation that he never understood – the fact that the cost for electricity will ultimately be the same in both costing structures. What Colbert is referring to is his learning about two different costing systems for electricity as described in the problem. In his view, both options end up costing the same. While Colbert's analysis of the task is mathematically incorrect, it is clear that the context of the task presented a real-life problem that, as he says, people in his neighbourhood have been arguing about. Looking at the graph below, illustrating the cost of electricity for the two options, it is clear that the two lines intersect at the point (800, 200), which means that if electricity consumption is more than 800 *kWh*, then the cost of electricity will be cheaper when using the first costing structure.

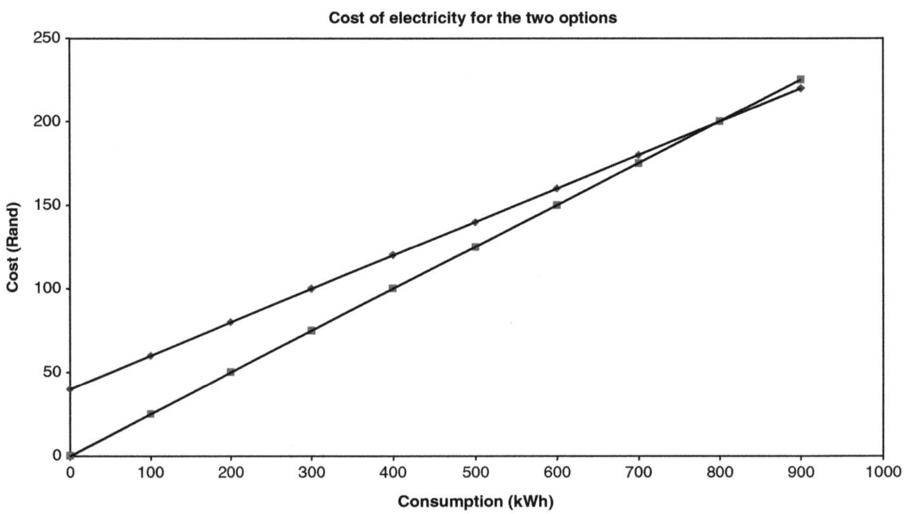

Given the learners' seeming reluctance to talk about the fact that their home languages were used, the interviewer asked them a direct question about the way in which their home languages were used in the task.

Interviewer:	I understand that the tasks that were given were written in both your home language and English. Tell me about that.
Sindiswa:	It was fine. It was just the same. It was the same as doing it in English, because I understand both languages.
Nhlanhla:	I think mem leyo kusinikeza amaphepha o i-two kuya nceda mem, ngoba, like mina, kukhona amanye ama-questions bengingawa-understandi, i-home language iyakhona ukusiza ukuthi ngiwa understande. **[Mam I think that one of giving us tasks in two languages is very helpful mam, because like there are questions that I did not understand and my home language helped me understand.]**
Colbert:	Iya, I think is a good idea, uyabona, ngoba iyenza ukuthi … iyenze izinto zibe simple, ngoba if singa-understendi ngeEnglish, sicheka ku … our languages, aba simple bese siyakhomphera. **[Yes, I think it is a good idea, you see it makes things simple because if we do not understand in English we check in our home languages and it is simple because we can compare.]**
Sipho:	Iya, I did understand in English and vernac. I did benefit.

From the learners' responses above, it is clear that none of them experienced the use of their home languages as a distracter or constraint. In fact, Nhlanhla and Colbert explained that having their home language versions was helpful. The silences and presences in the learner interviews are interesting. The interviewer explicitly had to raise the issue of language for the learners to talk about it. This suggests the transparency of language as a resource. Although the home languages were visible, in the sense that the learners were for the first time given written text during the mathematics lessons in their home languages, they are also invisible in that they were not distracting the learners' attention from the mathematics tasks they were doing. The learners were not focusing on the languages, but on the mathematics of the task.

As Lave and Wenger (1991) argue, for a resource to be useful it needs to be both visible and invisible. In their view, the invisibility is in the form of unproblematic interpretation and integration (of the artifact – in this case, the translated versions of the task) and visibility in the form of extended access to information. Although the unusual use of the learners' home languages in the task could be noticed and used, when invisible, it did not distract learners from the task. The learners were at liberty to choose which language version they wanted to refer to at any time. This contributes to the relevance of the solution being explored in this chapter. The learners can be given an opportunity to draw on the linguistic resources they have and, at the same time, the presence of English assures them of the fact that they are not losing access to the language of power, to which they so much want to gain access.

The above analysis suggests that viewing the various languages – home, English and mathematical – more synergistically and less as a dichotomy can lead to finding possible solutions for teaching and learning mathematics in multilingual classrooms.

Of importance is the fact that possible solutions are not just about language choice, but also about the nature of the mathematics tasks selected. The use of challenging but manageable mathematical tasks located in the learners' everyday context seems to have played an important role in bring synergy between the language of power (English) and other languages crucial for supporting conceptual mathematical understanding. The everyday contexts of tasks seem to arouse interest in the learners and also encourage the use of languages in a way that enables access to the mathematics register, rendering language an invisible resource for communication and discussion, rather than a visible obstacle. In this context, teachers use whatever language will aid the understanding of the mathematics task at hand. Furthermore, since the mathematics tasks that are located in the learners' everyday contexts are not algorithmic or formulaic, they are not from the teachers' own apprenticeship of observation in their own schooling; they are not predisposed to using only English as was the case when they were taught in school.

6 What Is It That We Still Do Not Know?

A recent review of research on multilingualism in mathematics education in South Africa highlights the need for more research in this area of study (Setati et al. 2009). This area of research is crucial, not only because it is important for equity and access for all to mathematics, but also because a majority of learners in South Africa learn in a language that is not their first, main or home language. Until these learners can have equal access to mathematical knowledge, it will be impossible to produce the number of engineers, technologists and scientists that South Africa so desperately needs.

Recent analysis, in which Kahn (2001) used language as a proxy, shows that it is mainly learners who are learning in a language that is not their home language who are not succeeding in grade 12 mathematics. As indicated earlier, poor performance by multilingual learners cannot be solely attributed to their limited proficiency in English. This attribution takes an individual deficit view of the learner instead of seeing learning as taking place in context. Research needs to identify other factors that interact with the fact that these learners have limited proficiency in the LoLT to contribute to their poor performance.

Research also needs to identify measures that can be taken to ensure success in multilingual mathematics classrooms. In this chapter, I have pointed out that this is likely to require the linking of cognitive and sociopolitical perspectives, as well as the deliberate, proactive and strategic use of the learners' home languages together with English in the teaching and learning of mathematics in multilingual classrooms.

Although we know for sure that language is important for learning and teaching, it is also crucial to improving learning and thus mathematics achievement. What we still do not know are the following:

- *Is the strategy of using learners' home languages in a deliberate, pro-active and strategic manner as explored in this chapter a solution that can lead to improved learner performance in mathematics (applicable to a wide range of mathematical learning activities)?*

- *What do all teachers need to know, and what skills do they need to develop, in order to be able to teach mathematics effectively in multilingual classrooms?*
- *What changes are required in mathematics teacher education to ensure that future teachers are adequately prepared to maximise the personal, linguistic and mathematical potential of all learners?*

These questions relate not only to issues of knowledge and expertise on the part of teachers, but also to issues of professional and personal identity. Much remains to be done.

Chapter 9
Discursive Demands and Equity in Second Language Mathematics Classrooms

Richard Barwell

Second language learners of mathematics face a double challenge: they must learn the language of the classroom and, at the same time, they must learn something of the mathematics that is presented, discussed and conceptualised in that same language. Second language learners include students from linguistic minority backgrounds whose home languages are not well-represented or recognised in wider society. Such learners are bilingual or, more often, multilingual, although their level of proficiency in any one language, or in some combination of languages, varies according to the situation and with what is being discussed.

A variety of terms or labels have been used to describe such students: for the most part, such terms originate in government policy. They include:

- learners who are Limited English Proficient (LEP) – in the U.S.A.;
- English language learners (ELLs) – more recently in the U.S.A. and Canada;
- learners of English as a second language (ESL) – in Canada and the U.S.A.;
- learners of English as an additional language (EAL) – in the U.K.;
- learners from non-English-speaking backgrounds (NESB) – in Australia.

These terms all come from countries that are portrayed as English-speaking and it may therefore seem unsurprising that they all take English as the reference language. However, these terms are descriptions of *learners*, not medical conditions. As such, they all index a deficit view of bilingualism or multilingualism, since they all highlight the value of English and leave students' 'other' languages largely invisible or inaudible.

Researchers have often argued for the use of alternative formulations, particularly 'bilingual learners', although such usage is complicated by the politicisation of bilingual education programs in parts of the U.S.A. (see, for example, Leung 2005).

R. Barwell (✉)
Faculty of Education, University of Ottawa, Ottawa, ON, Canada
e-mail: rbarwell@uottawa.ca

B. Herbel-Eisenmann et al. (eds.), *Equity in Discourse for Mathematics Education: Theories, Practices, and Policies*, Mathematics Education Library 55,
DOI 10.1007/978-94-007-2813-4_9, © Springer Science+Business Media B.V. 2012

In this chapter, I examine the challenges facing second language learners in mathematics classrooms. To avoid the kind of deficit assumptions already alluded to, I draw on a social, discursive perspective that sees second language learners' participation in mathematics classroom interaction as jointly achieved. In what follows, I set out the idea of *discursive demands* arising in mathematics classroom interaction. This notion is illustrated with data from a mathematics lesson in a multilingual classroom in London, U.K.

At primary school level, the teaching of mathematics in England is guided by the government's *Primary National Strategy*, as represented by a framework document and many other publications. References to the needs of EAL learners are widespread (often appearing alongside guidance on special educational needs) and are generally framed in terms of a metaphor of 'access' (Barwell 2004). The following statement, for example, is typical; similar statements appear in several parts of the framework:

> Children learning EAL must be supported to access curriculum content while also developing cognitive and academic language within whole-class, group and independent contexts. [...] it is critical to maintain a level of cognitive challenge consistent with that of the rest of the class. Children who are or have become conversationally fluent will continue to require explicit attention to the development of the academic language associated with the subject and of specific aspects within the subject. Planning should identify the language demands of the objectives and associated activities. Making sure that EAL learners know and can use the language demanded by the curriculum content of the unit or lesson then becomes an additional objective. To identify the language demands, teachers and practitioners will need to consider the language children will need to understand in order to access an activity. (DfES 2006, p. 14)

The access metaphor that is apparent in this statement constructs language as a kind of portal, through which students somehow must pass in order to enter a subject like mathematics. As I have written elsewhere (Barwell 2005a), this metaphor has several problematic aspects:

> First, language is separate from content, with the implication that if students can learn the language, learning the [curriculum] content will be straightforward. Indeed, it could also imply that language should be learnt *before* content. The idea that language is a part of content and vice versa is to some extent obscured. Secondly, therefore, the view of language as a portal renders language transparent, obscuring its role in the construction of a subject. Thirdly, both language and content are portrayed in rather static terms, external to the learner, obscuring the subjective experience and variable use of both language and subjects like mathematics. Finally, by obscuring the variability of language and content, their relationship with social, power-suffused relationships and structures is also obscured. Thus, for example, the political dimensions of language and the often authoritarian nature of school curricula are hidden. (p. 144)

The access metaphor is, therefore, problematic in its portrayal of both mathematics *and* language. This portrayal, furthermore, has implications for equity in relation to second language learners of mathematics. By downplaying the role of language in the construction of mathematics, an impression is created that mathematics is the one subject in which second language learners will have few problems – something that is certainly not the case for many such learners. And the presentation of both

English and mathematics as rather static entities serves to underplay the diverse conceptions or experiences of English and of mathematics that second language learners may bring.

The above DfES EAL statement also refers to the idea of 'language demands', which is, in turn, derived from a distinction between academic and conversational language based on the work of Cummins (e.g. 2000). This work is discussed below. At this stage, I will simply observe that, in the statement, language demands are related to access. Hence, it is recognised that a subject like mathematics involves some specific forms of language, but these forms are construed as part of the portal; they act, perhaps, as keys with which to open the portal that leads to mathematics.

Notwithstanding the problems relating to this access metaphor, it is worth asking what the nature of language demands might be in mathematics. Government documents tend to emphasise vocabulary (e.g. DfES 2000), although there is some recognition of other aspects of mathematical language (DfES 2002). Nevertheless, language demands in mathematics tend to be understood as clearly specialised vocabulary, grammar or syntax: language demands, then, at least in U.K. policy documents, are about the language system. Much less attention is given to the demands of mathematics classroom discourse – the broader ways of using language in talking about and writing about mathematics (see, for example, Barwell 2005b). The aim of this chapter is to introduce and illustrate the idea of 'discursive demands' as a way of thinking about some aspects of the double challenge faced by second language learners of mathematics. This idea combines concepts from bilingual education, particularly the work of Cummins, with a discursive perspective on mathematical thinking that foregrounds the situated, socially organised nature of cognitive processes like thinking, knowing or remembering. These ideas are discussed in the sections that follow.

1 Conversational and Academic Language Proficiency

Cummins' work (e.g. 2000) has been influential in shaping the direction of research in bilingual education, as well as in informing the development of pedagogic practices that are effective in supporting the school learning of bilingual students. One construct, in particular, has become widely used: the distinction between academic and conversational language, sometimes also referred to as Cognitive Academic Language Proficiency (CALP) and Basic Interpersonal Communicative Skills (BICS). This distinction initially emerged in research that sought to understand why apparently fluent bilingual school students were under-performing in tests (Cummins 2000, p. 58).

Cummins argued that a single construct of global language proficiency is insufficient to explain such students' under-performance. In effect, language proficiency is domain specific and, indeed, context specific. In particular, students may have a level of proficiency that allows them to participate fully in everyday conversation, while not having developed a similar level of proficiency in the academic language of subjects such as mathematics. Indeed, this point applies to all students to some degree:

> native-speakers of any language come to school at age five or so virtually fully competent users of their language. They have acquired the core grammar of their language and many of the sociolinguistic rules for using it appropriately in familiar social contexts. Yet, schools spend another 12 years (and considerable public funds) attempting to extend this basic linguistic repertoire into more specialised domains and functions of language. CALP or academic language proficiency [...] reflects the registers of language that children acquire in school and which they need to use effectively if they are to progress successfully through the grades. For example, knowing the conventions of different genres of writing (e.g. science reports, persuasive writing, etc.) and developing the ability to use these forms of expression effectively are essential for academic success. (p. 59)

It is notable that, for Cummins, academic language proficiency is closely tied to the language of schooling. That is, the construct 'academic language proficiency' is specific to the particular situation of schooling. Subsequent research (e.g. Thomas and Collier 1997) has confirmed the validity of Cummins' distinction and has demonstrated that second language learners take several more years to develop academic language proficiency, as compared with conversational language proficiency. A basic equity issue is immediately apparent, in that second language learners may be assumed to be 'fluent' in the classroom language – and treated as such – when they would, in fact, benefit from support in the development of academic language proficiency in subjects like mathematics.

Cummins (2000, pp. 67–68) goes on to refine the notion of academic language proficiency to take account of two different issues: situational aspects of language use and related cognitive aspects. To do so, he defines two inter-related continua. The first continuum extends from context-embedded to context-reduced communication. In context-embedded communication, interaction is supported by a wide range of situational or interpersonal cues. In a face-to-face discussion, for example, participants may draw on facial expressions, gestures, nods of the head and so on to make meaning, indicate comprehension, ask for clarification and generally communicate. In context-reduced situations, by contrast, the role of situation or interpersonal cues is greatly diminished, as, for example, in a formal written examination. Context-embedded interaction is typical of a great deal of everyday talk outside of school. Much of the interaction encountered *within* school is to a greater or lesser extent context reduced. Consider, for example, listening to a teacher's explanation, presenting a solution to a mathematics problem, writing out such a solution or taking a test.

Cummins' second continuum concerns the cognitive demands of interaction. Cognitively demanding interaction requires "active cognitive involvement" (p. 68), such as, for example, recalling and using new vocabulary or working with an unfamiliar genre or grammatical structure. Interaction becomes less demanding as it becomes, in effect, more automatic. He presents these two continua within a single framework.

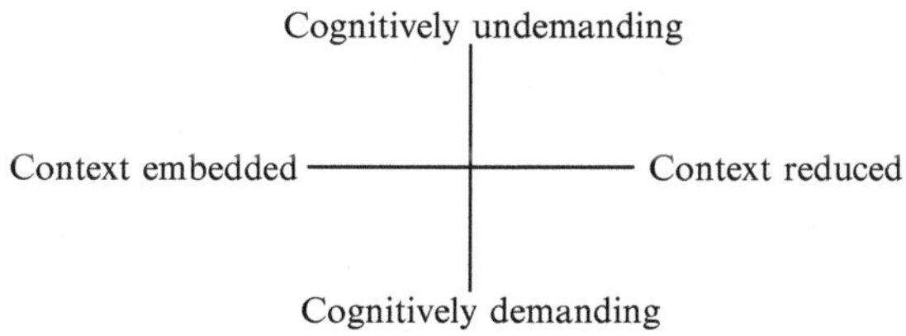

The two dimensions are highly interdependent. Face-to-face talk, for example, relies on a high degree of context in the form of gestures, facial expressions and the presence of many of the objects of discussion. Such context supports meaning-making and so tends to reduce the cognitive demands of the interaction. Some interaction involves more reduced contexts. Giving a presentation, for example, involves less direct interaction, so is more context reduced than is face-to-face interaction. A reduced context tends to lead to more cognitively demanding interaction – giving a presentation makes greater cognitive demands to produce appropriate language. Of course, what is cognitively demanding for one student can be relatively undemanding for another. In some sense, therefore, the framework can be seen as relative to the individual. Nevertheless, it allows for some broad general observations to be made. In particular, academic language tends to be both cognitively demanding and context reduced.

Cummins' ideas provide a valuable, though rather broad, framework with which to understand some key issues facing second language learners in school, as well as to inform teaching. While these ideas clearly recognise interaction as central to learning, they do not allow an examination of the detailed nature of this interaction. The framework is not designed with such a purpose in mind. Interaction between students or between teachers and students is also central to equity. It is therefore valuable to consider how the specific demands of interaction in mathematics are implicated in the participation of second language learners.

2 Discursive Demands: Theoretical Perspective

U.K. policy uses the term 'language demands', drawing explicitly on Cummins' notion of academic language proficiency. Both policy and to some extent Cummins tend to see these demands largely in terms of the language system, focusing on vocabulary, text genres, grammar, and so on. Research on bilingualism or second language learners in mathematics classrooms initially had a similar focus (e.g. Austin and Howson 1979). In recent years, however, researchers have emphasised how issues of vocabulary and grammar are only one, perhaps more salient, feature of learning mathematics in bilingual or second-language settings. Research by Khisty (1995), Moschkovich (2002, 2008) and Setati (2005a), as well as my own (e.g. Barwell 2009), all highlight broader discursive aspects of bilingual, multilingual or second language mathematics classrooms, including the use of multiple languages; the role of students' everyday language; the interpretation of graphs, tables and diagrams; the construction of students' relationships with each other; and political tensions surrounding language use. What this work suggests is that situated language use – i.e. discourse – in mathematics classrooms is as significant as the formal linguistic features of the mathematics register in second language learners' participation in and engagement with school mathematics. I propose to refer to these kinds of demands on second language learners as *discursive demands*.

The perspective I will use to examine the discursive demands of mathematics classroom interaction draws on discursive psychology (Edwards 1997, 2006; Edwards and Potter 1992; Wetherell 2007) and related ideas in conversation analysis (Sacks 1992). From this perspective, cognition, including mathematical thinking or language learning, is seen as situated, jointly produced, contingent and organised by the structures of interaction. Mathematical cognition (i.e. thinking, knowing, understanding, etc.) is constructed by participants through their interaction. What a student knows in mathematics is not simply a stable mental state, waiting to be produced at the appropriate moment. What a student knows is constructed through her or his participation in mathematics classroom interaction.

Discursive psychology, then, is less concerned with what students are 'really' thinking, in preference for a focus on how what students are thinking is portrayed and discursively constructed. The discursive construction of cognition depends upon some basic features of interaction, such as choice of words, descriptions and the structure of the talk itself. Indeed, the socially organised structure of talk is seen as more significant in meaning-making and the construction of cognition than the semantic content of the words used. For this chapter, I will focus on the following specific structures found in everyday talk: the role of turn taking and adjacency pairs; sequentiality; repairs; and recipient design.

Spoken interaction is typically structured in turns, with the *turn-taking* structure both enabling and organising interpretation. A common feature of turn-taking is the occurrence of two-part structures, such as question–answer, greeting–greeting or invitation–acceptance. These two-part exchanges are called *adjacency pairs*. The second part of an adjacency pair normatively appears directly after the first, hence the term 'adjacency'. In some circumstances, however, the second part may appear some turns later, often with other pairs nested in between, as in the following example, used by Sacks (1992, vol. 2, p. 529; see also Silverman 1998, p. 106):

A: Can I borrow your car?
B: When?
A: This afternoon
B: For how long?
A: A couple of hours
B: Okay.

In this exchange, the first and last turns in the extract form an adjacency pair, with two question–answer pairs inserted in between. An important feature of adjacency pairs is that once the first part has been deployed, it is difficult for the addressee to avoid completing the pair with the appropriate second part. Indeed, any response will be interpreted in the light of the adjacency pair structure. Even if, for example, B were silent after A's question, that silence would still be heard as a response – a possible refusal, for example. While the second part of the adjacency pair can be put off, as in the above example, it must generally be completed in some way. It is in this sense that interaction is fundamentally *sequential*.

This principle is about more than the basic sense of interaction unfolding over time; the sequentiality of talk is a part of its structure: a question requires an answer; a request requires an acceptance. Responses to first pair parts are interpreted in the light of the adjacency pair structure. Equally, and reflexively, the responses serve to construct the nature of the interlocutor's understanding of the first pair part. Where these understandings are at odds some kind of renegotiation arises, a process known as *repair*.

The purpose of repair sequences is to re-establish a shared sense of understanding, although 'understanding' here refers only to the explicit interpretations made available in participants' utterances, rather than any internal mental state. The principle of *recipient design* is simply that utterances are audibly shaped to suit whomever is listening. A sense of this is apparent in the above extract. Although no information is provided about A or B, the nature of their exchange suggests that they are well acquainted. The phrasing of the opening question 'can I borrow your car?' is familiar. There is no preliminary introduction of the topic and no reason is given. It might even be deduced that A has borrowed B's car before. Such inferences are possible because of the basic principle of recipient design. A's question is designed for someone he or she knows well; in this way, it also constructs the interlocutor as such a person. The subsequent turns in the extract are similarly designed. Recipient design is accomplished by various means, including choice of words, forms of address or by varying the amount of information given. (For summaries of the preceding ideas, see Silverman 1998, pp. 101–109; ten Have 1999, pp. 18–25.)

Conversation analysis is a form of micro-sociology that seeks to understand how social life is organised by participants. Discursive psychology draws on the assumptions and analytic tools of conversation analysis as a starting point for the examination of the social organisation of cognition. Basic structures of interaction, such as turn-taking, adjacency pairs and recipient design serve to shape the content of talk, including, for example, mathematical thinking or knowing. Such social structures are distinct from those of languages like English, although clearly they rely on specific linguistic structures. Hence, I will define discursive demands as the forms of interaction arising in classrooms through which second language learners, along with their interlocutors, jointly produce both cognition and context. In the next sections, I discuss excerpts from interaction involving a second language learner of mathematics during a single mathematics lesson. My purpose is to explore what kind of discursive demands might arise for second language learners of mathematics.

3 Discursive Demands in a Mathematics Classroom

K is a refugee Kosovan student. He joined his school in London, U.K., at the start of Reception (equivalent to senior kindergarten). There were 26 students in the class, including EAL learners from Kosovan, Bengali and both anglophone and francophone African backgrounds. K was assessed by the school as EAL stage 1 (new to English) in November. His teacher estimated that he was probably stage 2 (becoming familiar with English) at the time I visited his class. Such an assessment suggests that he was developing a reasonable level of conversational proficiency in English at that time. His parents were reported as being supportive, though K's mother did not speak much English. K had Albanian language books on English and mathematics. The teacher felt he had a good memory, citing spelling as an example, characterising his memory as "very visual". The teacher reported that K relied on guessing, often not listening to instructions before embarking on a course of action. The teacher believed K was working at a relatively high level in mathematics, but was concerned that he could not show what he knew. In school tests, he scored higher in English than in mathematics.

The lesson discussed in this chapter focused on halving and doubling. K was recorded throughout the lesson using an individual microphone. The lesson began with the students sitting together on a carpeted area responding to the teacher's introductory questions. Later, the teacher moved on to a problem-like scenario about two children who have various items, one child having double or half the amount of the other. One problem, for example, stated that one child had four cars and the other had double. The task was to work out how many wheels there would be for each of the two children involved. The teacher introduced the use of multi-link cubes formed into rods to support thinking about halving. Following the whole-class discussion, the students worked in pre-assigned groups on worksheets. K's worksheet included similar problems to those discussed earlier, including questions about cars and wheels. Another teacher (T2) joined the class for part of the lesson and supported individual students, including K, with their work.

Whole-class discussion in a lively year 1 (Reception) mathematics class is fast and furious, with many speakers often competing for attention and frequent side sequences in which students interact with each other. An utterance like 'I know', for example, can be seen as primarily a bid for the floor (i.e. the right to speak) rather than a definitive statement by a student about her or his mathematical thinking. For a student like K, the discussion presents a number of discursive demands, of which I will highlight three: multiple speaker interaction; frequent repair sequences; 'raising the stakes'. I will describe these demands in more detail and illustrate them with selected excerpts from the transcript – although the densely interwoven nature of these various strands throughout the lesson mean that what is presented is necessarily a simplification.

3.1 Multiple Speaker Interaction

The whole-class discussion in the lesson is characterised by rapid turn-taking exchanges involving multiple speakers. The teacher is generally the participant who nominates who may speak next and so manages the interaction. She also has rights to interrupt other speakers. Of course, there are many interruptions and speakings out of turn from the students, but these are deemed not legitimate, as indicated by such utterances being explicitly refused or ignored. In some cases, for example, the teacher tells students that she did not accept their response because they did not put up their hand and wait to be nominated. K must find a way to make sense of and participate in this kind of interaction.

In the sequence below, the teacher has introduced two characters, Charlie and Ben, whom the children have come across before. In reading the sequence, consider how it looks from the perspective of K, sat in the middle of the carpet, surrounded by his peers, with the teacher standing in front of them next to a small whiteboard on an easel. (Transcript conventions are explained at the end in Note 1.)

T Ben/ we've got Charlie and we've got Ben/ **now** Charlie is **six** years old
K? [I'm six!
T [should have/ should've called him K shouldn't I today/ but he's six years old/ and **Ben/** is **half/**[as old
S [seven
T no don't shout out/
S ^I know^
T he's **half/** as old/[as Charlie
S [one
T ^so quietly/[tell the person [next to you^
S [he's six [
S [three
T Charlie's six &
K? I'm six!
S [I'm six
T & [and Ben is **half** his age/ you're on the right lines R/ how can you use your [fingers to help you
K [it's three
T he's **half** his age
K seven
Ss six/ six
S two two two
S three/ three
T so if we've got six that's how many years/ old/ Charlie is/ so how old is Ben
Ss three/
Ss I know I know/
T I'm looking for someone putting their hand up really quietly/ K
K um/ three

T three/ how did you work it out?
K um/ [I just I just [I just thinked/ I just **thinked**
S [(I went like [this)
S [(…) [
T [no let him talk
what did you think?
K in my head **and** in my hands
T can you show me how did you think in your hands//
K then I just do **that**
T did you shall I show you what R did?

The teacher introduces the information that "Charlie is **six** years old" (lines 240–241), prefaced with an emphasised "**now**" to draw attention to it and an additional emphasis on six. This kind of presentation is typical of opening framings – it seeks to establish a starting point for subsequent discussion, in this case of a mathematical problem. K however, jumps in with 'I'm six!' (line 242), making relevant the fact that it is his birthday. The teacher acknowledges his contribution but shifts attention back to the mathematics problem she is still explaining. She repeats the information "six years old" and then adds new information that "**Ben/** is **half/** as old" (line 244). The emphases and pauses mark the shift from Charlie to Ben, as well as attending to the topically important information 'half'. There follows a variety of responses, both public and more private, with the teacher restating some of the information and managing the interaction in different ways.

Utterances that could be heard as mathematical solutions include seven, one, three, six and two. It is possible that some of these utterances are repeating information given by the teacher; for a student like K, however, what I want to highlight is the multiple responses in play, responses that he potentially needs to filter and evaluate. Furthermore, an exchange in the middle of the sequence illustrates another aspect of the demands related to multiple speakers: the neat adjacency pair structure, while still relevant, becomes somewhat problematic.

T Charlie's six &
K? I'm six!
S [I'm six
T & [and Ben is **half** his age/ you're on the right lines R/ how can you use your [fingers to help you
K [it's three
T he's **half** his age
K seven

While the teacher is, in effect, restating the problem, and appears to be addressing the student R, K is clearly responding to what she is saying. In particular, he says, apparently correctly, "it's three" (line 258), overlapping with the teacher. The teacher's next utterance is a repetition "he's **half** his age" (line 259), after which K says "seven". The adjacency pair principle means that, from K's perspective, the teachers repetition "he's **half** his age" can be heard as a response to his suggestion "it's three". Such a response, in classroom interaction, implies that the student's suggestion is incorrect. Alternatively, K might see that the teacher is directing "he's **half** his age" to someone else. Again, however, K might hear this as implying his own suggestion is incorrect. It may be, therefore, that K changes his answer, since the information he receives from the teacher, whether directed at him or not, seems to suggest his answer "it's three" is not correct. The point here is that the presence of multiple speakers makes it more difficult for K (or anyone else) to discern which utterances should be heard as relevant to their own contributions.

In the last part of the sequence shown above, K is nominated by the teacher and once more offers the response "three". The teacher accepts his response rather neutrally and initiates a question–answer adjacency pair:

T three/ how did you work it out?
K um/ [I just I just [I just thinked/ I just **thinked**

K's response to the teacher's question is 'troubled', meaning that it begins with a pause, and involves multiple repetitions. The nature of his response suggests that the question is in some (social, interactional) sense difficult to respond to. As with my earlier remark about the nature of 'I know', a statement like 'I thinked' can be seen as being more concerned with coming up with *some* kind of suitable account for where his answer came from, rather than being a specific description of a mental process. His struggle is compounded by several students' overlapping attempts to insert their own accounts in response to the teacher's question. While the teacher maintains her attention on K, eliciting the expanded account "in my head **and** in my hands" (line 276), the multiple speakers once again add to the discursive demands faced by K.

3.2 *Frequent Repair Sequences*

The phenomenon of repair is defined by ten Have (1999) as "organized ways of dealing with various kinds of trouble in the interaction's progress, such as problems of (mis-)hearing or understanding" (p. 116). He goes on to point out that repair is always initiated, for example, by responses like 'what did you say?' or 'I can't hear you' (see also Sacks, 1992, vol. 1, pp. 6–7). Repair sequences are likely to be common in classroom interaction. The following sequence arises after the teacher has asked the class how many wheels three cars would have. After various responses, one student, Rasool, makes an energetic contribution. Again, while reading the sequence, consider how it might seem from K's perspective:

Ras twelve twelve twelve! /[twelve/ twelve twelve/ twelve
T [you've got it on **two** cars
(*gasps*)/ how did you do that?
Ras ['cause/

S	oh oh/ I counted
Ras	I counted in my (…)
T	um Hakim I think you need to listen to Rasool/[‘cause I &
S	[it’s twelve
T	& didn’t get it as quick as Hakim did
Ras	counted on my fingers
T	right stand up show us/ stand up/ right how did you count on your fingers to get to twelve
Ras	because four add four makes t-t-twelve
T	oh does it
S	no eight
S	eight
T	four add four makes **eight/** so how many cars would that be
S	eight
K?	eight/ nine/
T	one car has four
K?	twelve!
S	four more!
T	four more would be?
S	eight
K	eight
S	twelve!
T	how many cars?
S	miss T
S	because I
T	not how **many** four cars
Ss	three!
T	no
Ss	eight!/ twelve
T	that’s one car/
S	two
T	quickly Zia
S	I’m thinking in my head
T	someone’s just said it/ Jane/
Jan	eight
T	no it wouldn’t be eight cars
S	four
T	right we’ve got three cars here/ how many wheels have we got on there?
Ss	four
T	four
Ss	four
T	so how many wheels have we got altogether?
Ss	[eight
Ss	[four
T	how many cars?
Ss	[eight

Ss [two
T how many **cars**?
Ss two
T two cars/ eight wheels
N I'm thinking in my head
T well you're very clever/ sit down/ three cars/ how many wheels?
N? twelve
S I can do all of those
T (*gasps*)// very **very** clever

In essence, the majority of this sequence is concerned with repair of Rasool's statement that "four add four makes t-t-twelve" (line 334). The need for repair is triggered by the teacher's response "oh does it" (line 335). The repair sequence, however, which continues for some time (until line 381) includes several embedded repair sequences. Rasool's initial explanation is, it turns out, problematic in two different ways. First, 'twelve' is not a suitable result for 'four add four'; second, and consequently, Rasool's explanation is not a satisfactory response to the teacher's question. The first form of trouble is solved immediately, with two different students supplying the solution 'eight', confirmed by the teacher. What then follows is a jointly produced repair of the second form of trouble – how to explain why 12 is the correct answer to the problem.

The repair begins with a question from the teacher: "so how many cars would that be?" (line 338). The sequence unfolds with a series of prompts and sub-questions from the teacher and a variety of mostly numerical responses from the students. Shared understanding is only re-established when the teacher asks, "how many wheels have we got on there?" (line 363) and receives the response 'four' which she accepts. It is only at this point that a degree of trouble has been resolved. This resolution seems to arise from the establishment of a joint focus of attention on a single car. The teacher's subsequent question, "so how many wheels have we got altogether?", confirms her interpretation of the students' preceding responses ("four") as referring to a single car. Her next question shifts attention to the number of cars (line 371) and, receiving as she does at least two different responses, a further repair is necessary. She restates the question with emphasis on "cars" (line 374). From here, a suitable explanation is completed in the form of a question–answer pair (lines 378–379).

It is apparent, then, that this sequence involves a good deal of trouble and repair, mostly initiated by the teacher's non-acceptance of some of the responses she hears from the students. Furthermore, these repair sequences are often layered, with sub-sequences repairing local trouble as part of larger-scale trouble arising from the request for an explanation for why there are 12 wheels on 3 cars. The structure is clearly rather complex and accounts for the sense that the discussion is not easy to follow. For K, then, keeping track of what is being repaired, including the different levels of embeddedness, represents another form of discursive demand.

3.3 *Raising the Stakes*

The last form of discursive demand I will highlight arises directly from the adjacency pair structure of interaction between two people. Specifically, the use of repeated questions raises the stakes for K. In the following extract, T2 is working with K and Steven, reviewing K's written responses on part of a worksheet.

T2 so you must write thirty two wheels/ and you too you've got to (…)// cross out your twelve/ how many does eight cars have/ how many wheels// thirty-two okay/
K I'm trying my second one//
Ste now you can do your **own** one//
T2 okay **now**/ four cars// d'you know what you've done look here// 'kay it's eight cars and it should be **double** eight and you've **halved** it/ you've made half of eight and it must be **double** eight/ what's double eight?
K umm=
T2 =eight plus eight
K two
T2 eight and eight together
K seven!
T2 what's eight/ and another eight/
Ste I know
T2 eight plus eight
K two!
T2 [no
Ste [sixteen
T2 sixteen
K oh
T2 so it should be sixteen cars/ /woah now you have to work out/ one and a six/

T2 indicates that K has mis-interpreted the question on the worksheet, saying that K has halved a number of cars, when the task is to double the quantity, thus triggering a repair sequence. She formulates this point twice, emphasising the words 'double' and 'halved'. She concludes with the question 'what's double eight?', which is contextualised by the preceding formulations. She has moved from interpreting the task to a direct question. By asking a question, the first part of an adjacency pair, she creates an opening for K to contribute, although the nature of the question also indicates the kind of responses that might be given: a number is expectable. K's response is 'umm', an utterance that allows him to take up his allotted turn, whilst buying some time. His turn is cut off, however, by T2, who reformulates 'double eight' as 'eight plus eight'. Such reformulations can be seen as guiding students, glossing previous utterances to provide a range of interpretations for the student to work with. They might also be seen as supporting the student in engaging with the language of the task, in this case by relating a mathematical term 'double' to an operation 'plus'. T2's glossing also serves to raise the stakes for K. Having been offered two formulations, 'double eight' and 'eight plus eight', there is a greater obligation on K to come up with a suitable response to complete the pair. This obligation, I should emphasise, comes from the interaction, rather than any intention on the part of the teacher.

It is a feature of talk that the more information that is provided with a question, the harder it is not to respond. K does provide a response: ‘two’. This response is generically suitable: it is a number. K has taken the turn for which T2 has nominated him, and rather than giving a non-committal ‘umm’, a response which was marked as unsuitable by the teacher’s swift intervention, K offers something generically appropriate which completes the pair. T2 again indicates this response is not suitable, however, by again reformulating, this time saying ‘eight and eight together’. The stakes continue to rise. K offers another generically appropriate but mathematically unsuitable response, this time as an exclamation, ‘seven!’ Again T2 indicates unsuitability by reformulating, “what’s eight/ and another eight” (line 690). This time Stephen takes the open slot, saying, “I know” (line 691). He indicates that the question is answerable and that, given the opportunity, he would be able to give a suitable response. The effect is to raise the stakes again. Not only is T2 reformulating the question, but Stephen claims to know the solution, implying K should too. T2 returns to an earlier reformulation ‘eight plus eight’ and K gives the same response he offered on the first occasion it was used: ‘two!’ Both T2 and Stephen break the pattern of the preceding turns. T2 now explicitly evaluates K’s latest (re-) offering, “no” (line 694). Stephen, overlapping, takes up the opportunity created by his previous turn, to give a response of his own, “sixteen” (line 695). This response is accepted by T2 through her repetition, “sixteen” (line 696). K accepts this closure, “oh”. Finally, the teacher recontextualises Stephen’s solution within the problem on the worksheet, by referring to ‘sixteen cars’.

Looking at the sequence as a whole, then, there are two features that place discursive demands on K. First, the interaction is structured by the question–answer format. Second, the sequence of reformulations, coupled with the adjacency pair structure, raises the stakes through the exchange. It is difficult for K not to respond, or to take too much time to respond, since the teacher’s questions expect answers. But the reformulations make it increasingly difficult for K to be wrong, hence raising the stakes.

4 Discursive Demands, Second Language Learning and Equity

I have described three forms of discursive demand that arise in the mathematical interactions in which K participates. The participation of multiple speakers in whole-class discussion is demanding, since it results in fast and furious exchanges in which several voices must be followed at once. It also results in ambiguity around the suitability or not of K’s own contributions. Is the teacher rejecting his comment or has she simply not heard it? Is her comment directed at him or at someone else? The frequent repair sequences, including embedded repairs, are demanding since, again, they must be tracked through sometimes lengthy exchanges. And these exchanges, of course, also feature multiple participants. Finally, in one-to-one interaction, extended question–answer sequences with reformulations of the questions can raise the stakes for K.

These *discursive* demands are different from *linguistic* demands as commonly understood. The linguistic demands of the lesson include the multiple formulations for 'double' and 'half', as well as the relating of these terms to various representations, including written symbols, cubes and cars. They also include the syntax of words like 'double' or 'half' – and note the contrast between 'double four' and 'half *of* eight'. The discursive demands I have discussed, however, arise from K's *participation* in mathematical discussion. While turn-taking, adjacency pairs and repair sequences are basic features of all spoken interaction, they are nevertheless relevant demands in K's participation in school mathematics. Any account of the potential challenges that K faces as he learns mathematics while also learning English cannot solely focus on the linguistic demands, important though they are.

How might the different forms of discursive demand described in this chapter interact with K's position as a learner of EAL? My first observation is that K is clearly able to participate in the question–answer pattern common throughout the lesson. He takes up turns when he is nominated, both in whole-class and one-to-one interaction. Indeed, the teacher's feeling that K is prone to 'guessing' can be seen as arising in response to this pattern. It may be less demanding to provide a 'guess' than to ask for more information or to find some other way out of the pattern, particularly when the teacher's reformulations raise the stakes or when other students are competing for the floor. Furthermore, K's responses are generically appropriate – they are numbers, for example – indicating more specific familiarity with the norms of mathematics classroom talk.

My second observation is that the range of formulations of 'double' and to some extent 'half', both in the whole-class discussion and in the one-to-one discussion, provide potentially valuable linguistic input, offering a range of ways of talking about this concept. In this particular sequence, K does not always appear to respond to these reformulations, but it may be that over time, he will become familiar with a number of ways of talking about 'double' and relate the concept to other arithmetic structures, including addition. It is noticeable, however, that throughout the lesson, K rarely uses the term 'double' himself. The occasions when he does so are in the form of repetitions. If meaningful production is an important part of the acquisition process (Swain 2000), however, whilst hearing various glosses for a term like 'double' is an important contribution to K's learning of the language of mathematics, supported opportunities to use such terms himself would also be beneficial.

At the start of this chapter, I argued that curriculum discourse concerning language and mathematics is based on an 'access' metaphor, which frames language as a portal through which students *get to* mathematics. From this perspective, K's task is fairly straightforward. To be able to learn mathematics, he needs to learn English in general and mathematical English in particular. He needs to learn words like 'double', 'half' and 'add' so that he can get to the underlying concepts.

The idea of discursive demands does not fit well with this model. Many of the demands faced by K will not be alleviated by somehow learning the word 'double'. It is not clear to me that he *could* learn what 'double' means *without* participating in the kind of complex, often challenging, interactions described in this chapter. These discursive demands are, in some respects, *prior* to the requirements to learn to use specific words like 'double'.

K's guessing, for example, can be seen as arising from the interactional patterns found in the mathematics classroom as much as from his arithmetic proficiency. The question–answer structure in the last extract in particular constructs K as guessing rather than as thinking or working out a solution. In the first sequence, K constructs himself as thinking 'in his head and in his hands', but the challenge of accounting for that thinking means that his thinking is not expanded into an acceptable explanation. In this sense, both exchanges may be seen as discursively demanding, despite being, in Cummins' terms, fairly context embedded. Furthermore, the use of reformulations, cubes, cars, and so on ostensibly serve to reduce the level of cognitive demand – although multiple glosses of 'double' might have the opposite effect. My point, however, is that while at the level of formal mathematical language K's task is to work out what doubling is and to do some himself, discursively there are other significant demands that arise from the structure of talk. If students like K are to be offered effective support in their learning of mathematics, this point should not be overlooked.

Note

1. Transcript conventions: Bold indicates emphasis. / is a pause <2 secs. // is a pause >2 secs. (…) indicates untranscribable. ? is for question intonation. ( ) for where transcription is uncertain. [for concurrent speech. ^ ^ encloses whispered or very quiet speech. = for latching (no gap between words). Italic capital letters indicate letter sounds: & indicates where turns continue on another line.

Chapter 10
A Discourse of Telling and Professional Equity

Beth Herbel-Eisenmann

Although the examples provided below come from interviews with people in different positions – Ruth, a tenth grade geometry student; Cara, a middle school mathematics teacher with almost 25 years of teaching experience; myself – I ask the reader to consider the similarities in the issues each of these people raise.

Ruth:	The first semester, it was more of … he [our teacher] would give us a problem and then tell us to apply it. And I was not used to that at all. I was not used to – in any class – being given a basic problem and then this, like, crazy, elaborate problem and saying, "You can use this equation to solve this problem." And I was, like, "I have no idea how to do that. You're giving me this huge problem and a teeny piece of it and you want me to, like…"
[Interviewer]:	Think?
Ruth:	Yeah, I don't know. Yeah, it is more of thinking than any other class just because usually, it's, like, [in other classes] this is a problem in the book, you get the same problem on your test with different numbers. But then [in this class] we're given more elaborate problems to solve and none of us are used to that.

(Ruth, in Obrycki[1] 2009, pp. 195–196)

> It was very helpful and very infuriating and frustrating at times because I wanted someone to just tell me what to do and not make me have to think. And yet that's exactly what I expect of my students, so I think that that was a good thing. And I loved the fact that I got to decide and find the point of things that I needed to change because we're not all the same. What I need to do differently is very different than what everyone else in this project needs to do

[1] In reference to items written by the teacher–researchers, I use their actual names with their permission to do so. In the transcripts, however, I use pseudonyms in order to maintain some confidentiality about some of the information they shared in the project. When we decided to write the book together, they knew that readers might be able to figure out who they were, but they agreed to contribute to the book anyway.

B. Herbel-Eisenmann (✉)
Department of Teacher Education, College of Education, Michigan State University,
East Lansing, MI, USA
e-mail: bhe@msu.edu

B. Herbel-Eisenmann et al. (eds.), *Equity in Discourse for Mathematics Education: Theories, Practices, and Policies*, Mathematics Education Library 55,
DOI 10.1007/978-94-007-2813-4_10, © Springer Science+Business Media B.V. 2012

> different or wants to do differently. So it had to be very open-ended and very non-judgmental and try to help us focus in and see what we need to do to learn and grow. It's one of the hallmarks of this project, but it's also been frustrating in that you know an easy answer would have been, "You tell me what I need to do and I'll do it," and it wasn't that way.
>
> (Cara, a project teacher)

> I've really been struggling this year [of the project] because I've never really facilitated action research before and I feel like there are things about what I value that I'm trying not to impose [by telling the teacher–researchers what I think they should do], but those things that I value make me ask questions that maybe other people in the group wouldn't. [...] And I'm learning a lot in the process, but I don't really always feel like what I'm doing just by asking them questions is enough. But I don't know what else to do.
>
> (Beth, in Adams 2009, p. 79)

In addition to the particular language choices articulated by Schleppegrell (this volume), there are other features of discourse practices that can help to investigate a range of equity issues. In this chapter, I explore a phenomenon that the people address in the quotations above: a "discourse of telling" in mathematics education.

In the first quotation above, a student named Ruth describes her reaction to her teacher giving her a "basic problem" and *telling* her to "apply" it to a "crazy, elaborate problem". She stated that this was unlike her other classes and it made her think in ways that other teachers had not – in those classes, she got "the same problem on [her] test with different numbers". Her teacher was involved in the professional development project I write about here. He was trying to help his students develop agency for their learning of mathematics by modifying his classroom discourse and investigating the impact of those modifications on his students' perspectives on their classroom experiences.

In the second quotation above, Cara, one of the other project teacher–researchers,[2] said that she sometimes felt infuriated and frustrated when I (as the organizer) did not *tell* her "what to do" in this same professional development project. She juxtaposed her own experiences and feelings with the fact that she expected the same from her students. Finally, I talk about not being satisfied with my practice as a facilitator and recognize that I purposefully was not telling the teacher–researchers what to do, because it seemed like it was imposing too much on their action-research decisions. In order to avoid a declarative "telling", I opted for an interrogative approach. I also recognized that my practice of "asking questions" was shaped by my values and wondered whether asking questions was "enough" to support the teacher–researchers in doing their work. My recognition of questions being related to my values, however, suggests that asking questions could be a another form of telling, albeit a more subtle one.

Contending with a *discourse of telling* was a parallel dilemma that occurred in each of two synchronous contexts in which the teacher–researchers and I worked for five years (2004–2009): the classrooms in which they were teaching students mathematics and the study-group community in which we discussed readings on classroom discourse and engaged in cycles of action research. Broadly speaking,

[2] I will use "teacher–researcher" and "university researcher", in order to distinguish the context in which we primarily worked and not as a way to privilege one over another.

the focus of the action research projects was to use what was learned about classroom discourse in order to become more purposeful about language choices in relationship to one's professed beliefs (see Herbel-Eisenmann 2010).

I identify a discourse of telling as a central dilemma to our project work, because we talked about it frequently throughout the duration of our project work and our discussions centred on both contexts (i.e. the classroom and the study group). As Gutiérrez (this volume) reminds us, attention to context is important because it serves as a humanizing tool in mathematics education research. In this chapter, I provide retrospective reflections about how exploring these two different contexts might have allowed us to examine the dilemma of telling in ways that could have opened up perspectives about the critical dimension of equity.

1 A Discourse of Telling

In order to address a discourse of telling, I begin by considering what has been written about telling in mathematics education. Because there are these different potential foci and viewpoints related to telling (as illustrated in the opening quotations), the phrase 'discourse of telling in mathematics education' is fairly ambiguous. It could be about the ways in which people (e.g. teachers, teacher educators, students, parents) tell or what these people say, when asked, about telling. In this chapter, I move among four of these foci/viewpoints: mathematics teachers' telling, mathematics teachers' talk about telling, mathematics teacher educators' telling and mathematics teacher educators' talk about telling. The dilemmas associated with a discourse of telling are related to equity and discourse because authority, control and power are central in decisions about when, how, why and in what ways one might decide to tell.

Descriptions of mathematics teachers' telling in classrooms are pervasive in many descriptions of classroom practice. Yet it is important to recognize that many of these descriptions did not draw on tools and concepts from discourse analysis as some other research has (see, for instance, Schleppegrell, this volume). Sometimes telling appears in the lectures that teachers prepare: the teacher maintains the floor for an extended period of time, explaining procedures for how to solve particular kinds of problems in the form of a 'monologue' (Lemke 1990). For example, Smith (1996) summarises literature on teaching as telling and describes the teacher's role: "provide clear, step-by-step demonstrations of each procedure, restate steps in response to student questions, provide adequate opportunities for students to practice the procedures and offer specific corrective support when necessary" (p. 390). The corrective support also typically includes additional explanation of the procedure.

Telling might also take the form of the familiar interaction pattern Initiate–Respond–Feedback (or IRF; see Sinclair and Coulthard 1975; Mehan 1979). In this pattern, what teachers initiate and how they provide feedback often tells students what is right or wrong in their thinking and controls who gets to talk and when. Both of these forms of telling are subsumed by what Nystrand (1997) calls a "recitation". He associates the epistemological stance of 'objectivism' with recitation, because he argues that an underlying assumption of this discourse practice is that knowledge is given; the focus is on transmission of knowledge from those who know to those who do not know.

With the increasing emphasis on constructivist learning theories, the kind of mathematics teacher telling described in the previous paragraph received much scrutiny. Although some interpretations of these learning theories suggested that teachers should not tell students anything, some researchers found this problematic. Chazan and Ball (1999), for instance, discussed the dilemma of telling with respect to the kind of classroom practices suggested in the U.S. *Standards* documents (NCTM 1989, 1991). The examples in the article, from Chazan's and Ball's own classrooms, provide images of mathematics teacher-telling as: making a comment or asking a question; introducing content that needed to be considered in an on-going disagreement; challenging students when statements were unclear or inadequate; opening up the discussion to additional participants; reintroducing ideas that were previously said that might reinvigorate the discussion and provoke reflection. These authors argued that such nuanced description can move us away from over-generalised notions of 'good' and 'bad' teaching, toward finer features and a position aimed less at evaluation: "We need to understand what kind of 'telling' it was, what motivated it and what the teacher thought telling would achieve" (p. 8). I return to this quotation later. It is important to note that this work marked a shift from telling being framed as a universally 'bad' pratice to reclaiming telling as a more contextually nuanced one.

More recently, Lobato et al. (2005) proposed that telling be re-named as 'initiating'. They showed that it included: describing a new concept; summarizing student work in a manner that inserts new information; providing information that students might need to test an idea or produce a counter-example; asking what students think of an idea; presenting a counter-example that might assist student thinking; engaging in Socratic dialogue to introduce a new concept; suggest a new representation. Finally, Baxter and Williams (2010) examined the discourse of telling and differentiated it from either *social scaffolding* (i.e. scaffolding that allows students to work together) or *analytic scaffolding* (i.e. scaffolding that is offered by tasks, teachers or one another, in order to build mathematical understanding). They found that some instances of social scaffolding involved a lecture in which the teachers clearly explicated how they wanted students to interact with each other, making explicit their expectations. When these expectations were not met, the teachers sometimes used questions to maintain or repair the students' understandings of the expectations. In analytic scaffolding, they found that teachers summarized discussions, related broader mathematical ideas to a task or offered an alternative strategy.

The literature reviewed so far only considered one of the contexts (i.e. mathematics classrooms) that I focus on here and equity concerns were not at the core of the authors' considerations. In fact, there was little or no mention of power dynamics among teachers and students, no consideration of issues about who gets to decide social and mathematical aspects of classroom work and little attention to control or authority[3] issues. Although we have some descriptions of what teacher telling looks like *in mathematics classrooms*, we do not have a parallel set of work that describes the discourse of telling *in the context of teacher education*. In other words, we have fewer images of what mathematics teachers say about telling, of teacher educators' telling or of what teacher educators say about telling. Thus, my focus here is related more to this latter context, but also focuses on the relationship between the context of teacher education and the context of mathematics classrooms.

In the literature related to teacher education, authors tend to focus on *teachers* 'telling stories' in professional development (see, for instance, Crespo 2006 or Nemirovsky et al. 2005) or on how teachers were working on when, how and why they told in their own classrooms (e.g. Baxter and Williams 2010). Descriptions of what teachers have to say about telling in their classroom teaching were not apparent in the literature I reviewed. As Herbel-Eisenmann et al. (2009) show, however, understanding how teachers talk about and make sense of discourse ideas in professional development settings can illuminate important issues and contextually nuanced aspects of discourse that might be absent in university researchers' descriptions. Thus, an important piece of work still to be done relates to exploring how teachers make sense of and talk about the telling that they do, so that we can support intentional use of this practice.

In research focusing on discourse patterns in professional development settings, there are glimpses of the kinds of things *teacher educators* might tell. For example, Chamberlin (2005) described a couple of instances of telling. The first was one in which the facilitator told teachers that they should take a few minutes to describe their students' thinking before they talked about it. The second was one in which the facilitator responded to a question about how she would handle an incorrect student solution by saying that she would not discredit student thinking but rather would try to understand the student's rationale for their thinking. In trying to understand the conversational patterns in her study groups, Crespo (2006) describes some of the moves she made, including encouraging norms of discourse and participation and trying to push the group's collective insights by means of "requesting elaboration, asking participants to comment on each other's accounts and asking others to comment on what made or did not make sense about what anyone said" (p. 48).

Crespo concluded her article by pointing out that we need to consider the extent to which work focusing on classroom practices carries over to adult learning contexts. There are other practices that have been explored more thoroughly across contexts: for example, the literature that suggests that teachers who engage in mathematical problem solving in professional development settings are more likely to teach problem solving well in their own classrooms. Yet, there are many other practices that might cross contextual boundaries and be relevant to developing practices. The work described in this chapter suggests that, even if we (as university researchers

[3] Chazan and Ball's article is an exception to this characterization because they did explore authority issues.

and teacher educators) do not cross these contextual boundaries in our examinations of classroom discourse, participants (in this case, teacher–researchers) do.

Similar to the literature about mathematics classrooms reviewed at the beginning of this section, I propose that we need to consider telling in the context of typical district-sponsored professional development (which might be more like the traditional teaching practices described by Smith 1996), as well as in collaborative professional development that is less common and long-term, such as study groups and action-research collaborations. In the same way that other authors in this volume argue that attention to discourse practices in classroom discourse can help work toward equitable classroom practices, I suggest that similar attention needs to be given in the work we do with prospective and practicing teachers. Like Chazan and Ball, I argue that in collaborations with teachers, teacher educators need to understand better what kind of telling we do, what motivates our telling and what we think it might achieve, as well as what teachers have to say about telling. In Sect. 3, in particular, I illustrate the discourse of telling that happened in the project work and, when possible, consider motivations and thoughts about what the telling might have achieved.

2 The Project Contexts

Although discourse has been the focus of research on mathematics classroom practices, there has been less attention to work related to engaging mathematics teachers in professional learning activities related to their classroom discourse. From 2004 to 2009, I was involved in a collaborative project with a group of eight middle grades (grades 6–10) mathematics teacher–researchers and a (now former) graduate student, in order to understand better how discourse literature might be helpful to secondary mathematics teachers. One long-term goal of the work involved supporting the teacher–researchers as they designed cycles of action research, in order to become more purposeful about mathematics classroom discourse. Thus, the project work was permeated by a focus on discourse, classroom discourse and mathematics classroom discourse. More recently, there has been increased attention to how discourse literatures might be useful to mathematics teacher's professional learning (see, for example, de Freitas and Zolkower 2009, 2011; Staples and Truxaw 2010).

I felt compelled to propose a professional learning experience related to classroom discourse that would be collaborative and in which participants felt they had a voice in the work. I decided study groups in which we would undertake readings and discussions together first and then support each other through cycles of action research fitted my beliefs and goals. Reading groups provided a venue in which we could learn about something we had a common interest in. Action research provided a venue in which the new ideas could be systematically tried out and investigated in practice. I felt fortunate to find eight secondary mathematics teachers who were interested.

In the first phase of the project, the teacher–researchers allowed the university researchers to collect classroom observations that were used as 'baseline' data, in order to consider what the teacher–researchers' classroom discourse practices were like, prior to engagement with readings and discussions about classroom discourse. The university researchers provided some written descriptions of these classroom observations, based on a first-pass, descriptive quantitative analysis, including, for example, how much time was spent on different 'activity structures', (using modified definitions offered by Lemke 1990). The teacher–researchers selected which activity structures they wanted to understand better and the university researchers undertook a detailed analysis and wrote analytic memos that described the discourse patterns we noticed in the selected activity structures. The university researchers drew on a combination of Systemic Functional Linguistics (drawing on, for example, Halliday and Matthiessen 2004; Lemke 1990; Morgan 1998; Schleppegrell 2004) and Critical Discourse Analysis (drawing on, for instance, Fairclough 1995, 2001; Hodge and Kress 1993) to write analytic memos for the teacher–researchers. The university researchers focused on how the teacher–researchers construed mathematics through language, as well as examined issues related to authority and control[4] (issues that the teacher–researchers discussed at length when they watched themselves on video). During the months that we did not videotape in classrooms, the group met and shared information about their previous teaching, the types of curriculum materials and tasks they used, and engaged in some activities in which we analyzed mathematical tasks or other artifacts of practice.

In the second phase of the project, we spent about a year as a reading group. Because I was interested in what the teacher–researchers would find compelling, I created a library of potential readings and organized them based on the scope of the discourse construct. For example, we read about words such as 'pronouns' (e.g. Rowland 1999), patterns of interaction such as Bauersfeld's description of 'focusing and funnelling' (e.g. Herbel-Eisenmann and Breyfogle 2005; Wood 1998) and larger sociocultural issues related to Discourse, such as Bourdieu's notion of habitus (e.g. Zevenbergen 2001b) or mathematics as a discursive practice (e.g. Adler 1999). The teacher–researchers selected which readings were of interest to them and we met weekly or bi-weekly to discuss them.

In the subsequent phase of the project work, the teacher–researchers selected aspects of their own classroom discourse to change and then carried out cycles of action research in which they studied the impact of the changes on students' social and mathematical experiences (see Herbel-Eisenmann and Cirillo 2009 for teacher–researchers' accounts of their action research projects). Prior to the action research cycles, each teacher–researcher created a visual mapping of what she or he felt was most important to his or her instructional decision-making. We referred to these artifacts as 'beliefs mappings', because we saw them as representations of 'professed beliefs' (see Gronewold 2009 and Lyddon Hatten 2009 for more about belief

[4] For some of our analyses involving authority, control and positioning, see Herbel-Eisenmann (2009), Herbel-Eisenmann et al. (2010), Herbel-Eisenmann and Wagner (2010) and Wagner and Herbel-Eisenmann (2008).

mappings). For example, the teacher–researchers wrote statements like, "Math is about thinking" and "Math should be intriguing to students". These beliefs mappings became the standards by which the teacher–researchers reflected on their own teaching. When they read any analyses that the university researchers had produced, when they watched themselves on video and when they asked for input from others in the group, the teacher–researchers typically used their beliefs mappings as standards for what they wanted to happen in their classrooms.

The design of the project created synchronous contexts in which the project occurred, including work being done in mathematics classrooms, as well as work being done in the project meetings. In this way, there were at least two synchronous contexts that were the focus of discussion in project meetings: the teacher–researchers' classrooms and their interactions with their students and the project meetings and our interactions. Unlike Gutiérrez (this volume), who focuses on rich description of specific contexts that were successful working toward equitable practices, I provide retrospective insights into how *overlapping* contexts may have prompted the teacher–researchers to consider the discourse of telling in ways they may not have otherwise.

3 Examples of the Discourse of Telling in the Project

I begin by illustrating, through project meeting transcripts, both how teacher–researchers talked about telling and about my own practice of telling in this project. I first offer an extended transcript from a discussion at the end of the first year of the project. Because we had recently completed collecting the fourth week of baseline data in the form of classroom observations, this was a transition time into the study group context in which we would read professional literature. I pointed out that we would be undertaking some discourse analysis of the classroom observation data and would, eventually, explore some of the patterns that we noticed in the classroom practice, too. This information, along with what they learned in the readings, could help them decide which aspects of their practice they thought they needed to do differently. My goal, I explained, was to provide them with as many possible patterns as we noticed and to get their input about what was most interesting, useful or compelling to them. I also wanted them to know that it was up to them to decide what needed to be changed and/or what might be different.

Beth: And then after [we describe as many discourse patterns as we notice in the analysis and we have finished discussing the readings you select] it's up to you to decide which things you like or don't like or which are things where you say, "This isn't such a bad thing but I know I can make it better if I do these things". And then this is where it's sort of amorphous [...] Where you're starting to decide and take a little more control over what you're doing in more conscious ways than maybe before.

Kate: Beth, you're always very good about being so respectful of what we do and what we think that you don't insert your opinion very often. One of the things that I would like is if we're looking at certain patterns and you know this is not a good pattern, you're not gonna let us, like, decide ourselves. I mean you're gonna help us know something is not a good pattern

(An eruption of overlapping talk from many people in the group.)

Beth: Well, in some cases though the things that catch my attention as a researcher are things that might be problematic or things that could be powerful. Just as a really basic example, as we're doing some of the analysis, some people are using words like *this*, *that*, *it*, instead of really telling the kids what the *this*, *that*, *it* refer to. And so by the time that you get two-thirds of the way through the lesson, you're like, "What is this thing?" So some of it might be deciding, "Oh, I'm using *this*, *that*, *it* too much. I need [to be more mathematically precise in my language use]. I'm talking about this person's strategy and this is what the strategy is," and recording it so there's some record up there that the kids are seeing what the *this*, *that*, *it* is. So it could be that we would describe a pattern like that.

Helene: So you wouldn't let us, just not see that. You'd help us see that?

Beth: Yeah. We'll be describing the patterns we see.

Stacey: Yeah, because when you made the comment a minute ago, "And then it'll be up to you guys to decide", I'm going [makes a face that indicates she feels panicked].

Beth: Well, but see that's the power of action research. You get to decide what you are or are not happy with. [...]

Cara: But you also have to remember that we all said that we're here because we want to grow and learn and so we will take it in that spirit. [...] But many times you do it and you don't even realize that you've done it. I used to say, okay, take out your math book. Okay, get out your pencil. Okay, okay, okay, until somebody pointed out to me that I was saying 'okay' fourteen times in a two-minute session. I didn't know and I had to really concentrate on getting that out of my teaching because that gets redundant.

Beth: So yeah. Some of that kind of thing will [appear in the patterns we describe]

Stacey: [...] Well, you know we are all having anxiety over that and you're probably right (several people talking) [...] It's just not knowing what you'll find that's so scary.

Beth: Right. And at this point we couldn't even tell you what they are cause we're still looking at the [data].

Stacey: Or what if there's so many that are obvious? That's where I'm at. It's like there's gonna be so many that, how am I gonna know which one [to focus on]?

This excerpt of transcript highlights the teacher–researchers' recognition of my discourse of telling by pointing out that I often chose not to "insert [my] opinion". As Kate so aptly pointed out, what I did *not* tell them was whether I thought their specific patterns were 'good' ones or not. For Stacey, this created a feeling of panic or anxiety. Cara agreed and reminded me that they chose to be involved in the project because they wanted to 'grow and learn'. Cara also pointed out how some language practices are beyond consciousness – for example, her repetitive use of 'okay' until someone brought the routine to her attention. Her example, however, was not related to the construal of content or positioning that Schleppegrell (this volume) described. Rather, it seemed to be related to the distraction of a redundant, unnecessary routine.

In this discussion, there is a potential that the issue the teacher–researchers were raising was not related to whether they were growing and learning or not. Rather, the issue may have been related to the fact that *what* I was telling them was quite different from what they were used to being told in professional development contexts. For example, I did tell them what I noticed about their classroom discourse by describing patterns that I observed in an analytic memo. I provided an example of something that I thought some of them might be dissatisfied with (i.e. repetitive use of vague references for mathematical objects and processes). I also told them that they had to decide what they liked or did not like about their discourse patterns, that they had to decide what to focus on and that they could figure out what might work better. It may have been that Stacey's statement about "having anxiety" over not "knowing what [we will] find" was also related to the fact that we would not be telling her whether the findings were "good" or "bad".

Throughout the transcript, my motivations for telling in these ways also appear explicitly. I acknowledge that discourse practices are not necessarily conscious. I decided to offer descriptions of the discourse patterns, in order to help raise awareness of current discourse practices. My understanding of facilitating action research groups underlies my points about wanting them to decide which discourse practices they want to change or what discourse patterns they were not happy with. As my opening quotation suggests, however, I was struggling with this aspect of our work. Although I brought a critical perspective to our work, as a consequence of this perspective I tried not to impose my own agenda on them. Consciously, I reverted to asking them questions because, at the time, I did not see some of my contributions as 'telling' as I now do.

Almost every time we talked about my practice of telling in relation to the professional development work, one of the teacher–researchers connected it to his or her own students. Talking about my telling in the professional development context, in fact, seemed to provide a bridge to talking about the teacher–researchers' telling. Often when they talked about telling in their own practices, they discussed issues of control with which they themselves were contending or about how students were resistant or became angry when they did not meet the expectation of telling them things they thought the teachers should tell. (See Obrycki 2009 for a discussion of this issue with respect to Ruth's opening quotation.) There were other aspects about changing the discourse of telling, too, that were raised later in the project work – for example, as seen in the following interaction that occurred in the third year.

Kate: But there are points [in time] when I just think, I mean, more than once I've thought, "God, I just wish Beth would just tell me what to do instead of making me go through all this stuff".

BHE: But I'm hoping you know by now that that's not going to happen.

Kate: I'm catching on. [laughter]

Cara: I also think about that with our students. I have them doing all these investigations and they are thinking, "Just tell me what I have to do. Do I add or subtract?" [inaudible]

Stacey: The kid that always says, "You're not ever going to tell me how to do this, are you?" I'm like, "No. But I'm going to help you get there." But I mean, just that realization washes over them and like, "Oh, God."

Helene: That's still more sophisticated than, "You're never going to teach us anything, are you?" And that's the message that gets sent home.

BHE: But part of it for me is I can't define [what you should do] because it's based on your goals. I think the whole thing about what it is you want to happen, and then trying to figure out what things will help you and your students do that, isn't something I can tell you.

Kate began this excerpt by focusing on my telling, stating that she thought "more than once" it would be easier if I would just tell her what to do rather than "making [her] go through all this stuff". This expressed her discomfort and in some ways served as evidence that, although she was one of the most experienced teachers in the group who had won awards for her teaching, making decisions about how to improve her own discourse practices was difficult. I reminded her that I did not think it was my role to tell her what to do. Later, I explained my motivations for this: I made this decision because I believed that only they could figure out what they needed to do, because they knew their students and context in ways that I could never know and that this kind of knowledge would allow them to do what was best for their students. On other occasions, I told them that they lived their practice every day and I could only visit their classrooms for short periods of time, so I trusted their ongoing relationship with students to figure out what they needed to do differently.

Cara's contribution connected our discussion about my telling with her own struggles with the discourse of telling in her classroom. She highlighted how she asked students to investigate mathematics, while they mainly wanted her to tell them what to do. Stacey animated a common conversation she had with students, in which she told them that she would not tell them how to do things but would "help [them] get there". Helene turned the conversation away from students wanting the teacher to tell and introduced another issue related to telling in mathematics classrooms. Her reference to "You're never going to teach us anything, are you?" illustrated her concern for students and parents interpreting her not telling as her not teaching anything.

In some ways, talking about my telling allowed the teacher–researchers to reflect on their telling in their classrooms. As these two project discussion transcripts highlight, the teacher–researchers had many concerns about my telling, as well as telling in their own classrooms. They were worried that they might have been telling when they should not have, about students' reactions toward them when they chose not to tell and how parents interpreted students' reports about them not telling. I elaborate these concerns in the next section.

4 A Discourse of Telling and Equity

When we consider the teacher–researchers' discussions about their discourse of telling, the motivations for their concerns make sense within the broader policy context of public schooling in the U.S. Considering teacher telling in mathematics classrooms, there is a pervasive belief that the teacher *should* be telling students what to do and how to do it in detailed ways. Doing otherwise opens a potential for turmoil, because a different form of telling is not what students or parents might expect, given their previous school encounters. Ruth's opening quotation, in which she describes her experiences, highlights this sentiment.

With almost daily reminders of accountability and pressures from students and parents, it seemed as if the teacher–researchers felt pushed toward attending primarily to the aspects of Access and Achievement or the 'dominant axis' of equity (Gutiérrez, this volume) when talking about their telling. The accountability measures that their students would face forced the teacher–researchers to attend to a long checklist of standards and to focus on whether they were getting through the list so students had exposure to the ideas that would show up on standardized tests. In some of the schools, the teacher–researchers worked to stay off the list of "schools in need of assistance", a designation which had strong repercussions with respect to the *No Child Left Behind* legislation. For example, if a school received this designation too many years in a row, the school could be closed, students would be sent to other schools and teachers could lose their jobs. These policies were very real for the teacher–researchers.

Yet, as the teacher–researchers continued to talk about the discourse of telling in the professional development context, other issues appeared that were not related to the dominant axis. For example, in a response to Whitenack and Yackel's (2002) article on the importance of teacher support when students are learning to justify and explain their thinking for the purpose of making mathematical arguments, Stacey began the discussion by considering the language moves the teacher in the article made:

Stacey: And I think what I do is I think I'm trying to facilitate the discussion but what I do often times is I get in the middle of it and I try to control it too much. And that's not the way my nature is at all so I just don't know why I do that.

Jeremy: Now, see, I was thinking the same thing. The reason I jump in sometimes is totally for control. And I would assume that's what all of us are doing.

Jackie: Yesterday I had set it up we were doing this activity and it was gonna be sort of like a group quiz. And I was going to randomly call on a kid [using the random generator on my graphing calculator] so they would all be accountable. [...] But then I designated, I had them elect three coaches. So if whoever went up was uncertain or had a question or something they could invite the coach to help them, one of the coaches. So the kids, I mean the coach was doing a great job but the kids were doing a really great job of asking the coach an appropriate question. And I started answering things. [And I started thinking to myself] "You know wait a second. You elected a coach for this job who is doing a great job, shut your mouth." It's really hard.

Cara: That's just because we've been so used to doing that.

Jackie: Yeah and sometimes I'm not sure that it's clear enough for everybody or whatever. Well, you know, they could ask their question [to the designated coaches]!

[a few turns deleted – teacher–researchers talking about the ways in which they ask questions]

Beth: So I want to ask you a follow up question about the things you were just saying about control. Do you think that you might ever do that, not for control but for other reasons?

Jeremy: Um, to jump into the argument? Oh, I would think when you feel like the discussion is going the wrong way and you're not comfortable with letting the kids have incorrect misconceptions up there. Or all of a sudden you start feeling like they're going a different direction and you want – you know, "What I need to teach you to do it is this way". I need them to go this way so I'm going to jump in and it doesn't really matter. So again, I don't really think there's any right or wrong. It's just all improvising. It's all jazz. Um, but I think clearly my tendencies, especially last year, were to do it more for control. And this year I am making more of a conscious effort to do it for those reasons: to clarify and to expand when I need to or when the conversation is totally faltering.

This discussion highlights several issues about dilemmas associated with how teachers think about telling. In particular, it highlights how the teacher–researchers grappled with authority and control, two issues to which telling relate. I first drew attention to the teachers' professional engagement with issues raised about their communication practices. Paying close attention to the way they used language and exercised authority had the potential for them to become overly nervous about how they interact with students; but instead, these teacher–researchers seemed to be becoming more careful in their consideration of their practices. As they became aware of their practices of unnecessary control, they also considered a range of the realties they faced. They thoughtfully considered good reasons for taking control.

They highlighted in the conversation excerpted above, for example, how they might tell more when the discussion was not mathematically productive or when they thought more needed to be clarified. Jeremy talked about how, prior to the project work, he used telling to control much of what was happening in his classroom. He felt that it allowed him to minimize potentially unproductive social interactions or misbehaviour. Now his practice of telling was more nuanced and his decisions were more purposeful. The teacher–researchers were trying to concentrate on making their reasons for telling and controlling more mathematical rather than social, a tension that Nathan and Knuth (2003) also described in their work related to classroom discourse with a middle school mathematics teacher. The teacher–researchers also wanted to create a safe space in which their students could explore mathematics, but they also knew that only focusing on the creation of a community did not always result in powerful learning. The teacher–researchers wanted to position their students as mathematical thinkers and knowers.

The teacher–researchers' long-term engagement with study groups, and with action-research cycles related to classroom discourse more generally, seemed to provide opportunities for them to shift from focusing primarily on Access and Achievement to considering issues related to Power and Identity. For example, after reading Zevenbergen (2001a), one teacher–researcher began to explore how his experiences were different from his students, allowing him to understand better his own primary and secondary Discourses and to question why he controlled aspects of his classroom discourse (Marks 2009). Another teacher–researcher began to see her students in more nuanced ways, in terms of how they interacted with one another in small groups, a setting that she had not considered prior to her videotaping her classroom (Lyddon Hatten 2009). For example, she observed a student who often struggled in class avoid talking about mathematics in a small-group interaction, while a student who had high academic status in the class took over the mathematical work, discounting some of the other student's contributions. These observations allowed her insights into both mathematical and social positioning of her students.

In contrast, my own discourse of telling in the professional development context distantly related to the components of the dominant axis (Access and Achievement). All of the teacher–researchers had access to basically the same resources, like readings and funds to attend professional conferences throughout the project. Another resource that was made available to them was individual attention: it allowed us, for instance, to provide every teacher–researcher with descriptions of her or his classroom discourse and to support them as they made changes and systematically studied them. It is important to point out, however, that these kinds of resources were not available to most of their colleagues and sometimes caused issues in their working relationships. In relation to Achievement, 'outcomes' were articulated by the teacher–researchers when they created their belief mappings, rather than determined by someone other than themselves. No measurements were used except when the teacher–researchers decided to employ them for their action research projects and often these were short surveys, journal responses or interviews that were carried out with students. In fact, the teacher–researchers worked collaboratively to develop an observation instrument that they could use to examine their practices, based on a set of ideas around which the group coalesced. These qualitative measurements allowed the teacher–researchers to pay closer attention to their students and were treated as evidence of students' experiences and perceptions. Also, because we did not have someone else's deadlines to meet, our time-frame and work were less structured and driven by time-related deadlines like assessments.

My choice to tell differently by not evaluating or labelling their practices as good or bad was related to my sensitivity to the teacher–researcher's developing identities as teachers and people. Through the project work, their awarenesses were raised and their identities were constantly being called into question. Because of this concern, the professional development context allowed Identity and Power issues, components of the 'critical axis', to be foregrounded. Many of the teacher–researchers described the process of seeing their subconscious discourse patterns and of watching themselves on videotape as 'painful'. In fact, some of the teacher–researchers became more aware of the ways in which their students had to downplay personal, cultural and linguistic capacities, but they were not always sure what it meant for changes in their practice. Awareness came first; knowing what to do and making advances that are sensitive to the students took much longer and was significantly harder. Their discussions about these realizations made them more aware of the ways in which their tacit discourse practices may have contributed to the marginalization of students and the ways in which the institution of schooling and people in the community may have perpetuated inequities.

By not being in a position to have to evaluate their practices, it was easier for me to attend to the teacher–researchers' needs both individually and as a small group. I focused on face-saving within the group and moderating discussions that called into question the teacher–researchers' personal, cultural and linguistic capacities.

Although we did not focus directly on our own privilege in society, we did grapple with how often our own experiences and discourse communities might be different from those of our students and what those differences might mean to classroom practices. If I had worked with a much larger group, it would have been harder to attend to the ways in which these realizations impacted them as people. Yet, as I expressed in my opening quotation, I was not sure that the things that I was doing were always the most helpful to the teacher–researchers as they contemplated, changed and investigated their practices. In fact, a recent investigation has led me to re-consider some of the assumptions I held coming to this work (see Males et al. 2010). Additional investigations of the discourse of telling could help guide other professional development work in terms of equity, in particular.

5 A Rumination on Context

I opened this chapter by stating that attending to context was important and, throughout the chapter, I have addressed the professional development context and the classroom context as if they were separate. There are, in fact, some important and obvious differences between the two contexts: for example, the teachers were working with children and I was working with adults; the teachers were in the group because they chose to be, but their students were in school because they had to be; the teachers did not have control over the issues of time they were allocated with students every day, while our time together was mutually decided (both in terms of when we would meet and for how long); the teachers worked within a tracked and age-defined system, whereas our group consisted of a range of levels of experience and expertise, as well as variation in the number of years of teaching experience; the teachers worked with about 150 students every day, and these students changed every year, while our group was almost constant over the 5 years of the project (11 for the first 2 years, 10 thereafter).

I would be remiss, however, if my separation of context led the reader to think that these contexts are disjoint. Cara's opening quotation draws me back to a crucial point: experiences in one context often and, at times, necessarily influence experiences in other contexts. An important reason to consider carefully the relationships between and among contexts is that the teacher–researchers often did when they discussed the discourse of telling. In fact, there are probably other discourse practices in mathematics classrooms that teachers also interpret and consider in terms of other contexts – for example, their lives outside of the classroom (see Herbel-Eisenmann and Wagner 2010). Yet, we know little about the ways the various contexts in which teachers live might influence the ways in which they consider discourse and the dimensions of equity described by Gutiérrez (this volume). We also know little about how to capitalize on the connections between and among contexts to help teachers consider their classroom discourse with respect to all four dimensions of equity, work that is crucial to equity in mathematics education.

Acknowledgement I would like to thank the teachers for collaborating with us, Michelle Cirillo for her contributions to the work and David Pimm, Dave Wagner and Jeff Choppin for insightful discussions and valuable feedback related to this chapter. This study was supported by an NSF grant (#0347906, Herbel-Eisenmann, PI). Any opinions, findings and conclusions or recommendations expressed in this chapter are those of the author and do not necessarily reflect the views of the NSF.

Chapter 11
Studying Discourse Implies Studying Equity

Candia Morgan

Within the mathematics education literature, the term *discourse* is used with a broad range of meanings, from referring simply to any communicative activity to referring to configurations of particular ways of speaking, behaving and viewing the world that structure participation in specific forms of social activity, drawing on the Foucauldian notion of 'orders of discourse' (Foucault 1972). (The latter type of usage may be distinguished by Gee's (1996) use of the capitalised *Discourse*.) Fairclough (2003, p. 61) provides what I consider to be a minimal conception of discourse as the linguistic[1] moment of a social practice (the other moments being material activity, social processes and relationships, and mental phenomena). The important point from this conception that I take to distinguish *discourse* from mere *language* or *communication* is that it is understood to be inextricably embedded within practice – not just use of a sign system or even an interaction between two or more individuals, but an interaction that has significance in some sphere of human activity, involving some form of relationship or relationships between the participants. This inextricable connection to social practice inevitably raises issues of equity when studying discourse. Discourse is not just about exchange of information but also involves establishing, negotiating and maintaining relationships among its participants and ways of understanding the world and our experience of it. It is in these relationships and world-views that inequities become manifest.

Understanding discourse in this way, then, entails that studying discourse is not simply a branch of linguistics, but can also be seen as a form of social research.

[1] The term *linguistic* needs to be interpreted very broadly to include at least visual and kinetic modes such as diagrams, graphs, gestures, etc. as well as language (Kress and van Leeuwen 2001). Certainly, in mathematics, such diverse and specialised modes play a major part in allowing the construction of mathematical meanings.

C. Morgan (✉)
Institute of Education, University of London, London, UK
e-mail: c.morgan@ioe.ac.uk

B. Herbel-Eisenmann et al. (eds.), *Equity in Discourse for Mathematics Education: Theories, Practices, and Policies*, Mathematics Education Library 55,
DOI 10.1007/978-94-007-2813-4_11, © Springer Science+Business Media B.V. 2012

As such, it is of enormous relevance in the field of education. Over and above the role that language and interaction play in learning – for example, in the move from the interpersonal to the intrapersonal plane posited by Vygotskian theory (Vygotsky 1986) – we must consider how the discursive moment of educational practice contributes to the socialisation functions of education, including the formation of identities and the production and reproduction of social structures. In so doing, I draw on Fairclough's Critical Discourse approach (Chouliaraki and Fairclough 1999; Fairclough 1995, 2003), which demands that, when analysing any text, the analyst should pay attention to the language of the text itself, the role that it plays in the social practice of which it is a part and the social structures in which this practice is embedded.

The chapters in this part each address more than one of these levels of analysis, though they locate their emphases differently. Apart from Setati's chapter, however, they do not substantially address the level of social structures; rather, they privilege the analyses of language and social practices. Of course, when dealing in a detailed way with relatively small quantities of qualitative data arising within a single setting, the higher-level social structures and their effects are not easy to distinguish. However, I would argue that, for the broad programme of research concerned with discourse and equity, it is necessary to take such structures into account in order to be aware of the ways in which phenomena apparent within a particular social practice may arise from or have impact upon the lives of the participants beyond that practice and to be able to consider the possibilities of more equitable practices.

Schleppegrell's chapter focuses primarily at the level of texts – specific instances of interaction. Drawing on Halliday's Systemic Functional Linguistics, she offers analytic tools that allow us to consider how the nature of mathematics is construed in classroom texts and how students and teachers are positioned by the linguistic forms of interactions. A key insight of Halliday's social semiotics is the recognition that every communicative act involves field (what the communication is about), tenor (the social situation in which the communication takes place) and mode (the channel or means of communication) and that these three aspects are realised in the ideational, interpersonal and textual functions of the communication. Making use of his grammatical tools thus enables one to engage in linguistic analysis, but also entails interpretation of such analysis within the context of the communicative act.

Schleppegrell outlines how the grammar of the linguistic component of communication allows us to unpick the ways in which speakers' and writers' linguistic choices construe ideational, interpersonal and textual meanings, identifying how utilising such analytic tools may be used to investigate equity issues. Focusing at the level of the communicative act in the classroom or even at the more general level of on-going classroom practice, however, does not allow us to engage fully with two fundamental questions: *Why do teachers and textbook writers speak and write in ways that construe particular views of mathematics and particular positionings for their students? Which students are advantaged or disadvantaged by the spoken and written texts in classrooms?* These two questions demand that we look outside the immediate practice of individual classrooms to consider the dominant discourses and social structures that shape the resources that students and teachers bring with them into the classroom.

Also focusing primarily at the level of the text, Barwell provides an analysis of a specific instance of classroom interaction. Though his analytic tools address the interaction rather than the clause, his analysis operates at the same level as that proposed by Schleppegrell, demonstrating how student 'K', at an early stage of learning English, comes to be positioned as "guessing" in his mathematics class. At the beginning of his chapter, Barwell sets the scene by considering some relevant aspects of the broader social structure, indicating the status of students from linguistic minority backgrounds and providing some insight into the policies and dominant discourses that affect how education is provided for such students and the perceived relationships between mathematical content and language. This begins to allow us to address the questions raised in the previous paragraph, though there is still work to be done to establish the routes by which these (and other) structural aspects, together with the on-going local classroom practice, lead to the particular form of questioning found in this classroom interaction.

In the professional development programme described in Herbel-Eisenmann's chapter, the participants use analysis of instances of interaction as a means of investigating and questioning their classroom practices, implementing Schleppegrell's proposal that text-level analysis is a tool for both researchers and teachers to investigate equity issues. However, Herbel-Eisenmann's primary focus is at the level of social practice: the 'dilemma of telling' she explores is a component of the practices of both school classrooms and teacher professional development programmes. Again, she provides some discussion at the structural level of policy and administrative factors that shape 'usual' practice. This discussion touches on the first question of why 'telling' is so common in classroom practice, though it does not address the question of which students might be advantaged or disadvantaged by this.

The programme that forms the context of Herbel-Eisenmann's study may be considered an 'alternative' form of practice, attempting resistance to dominant discourses of teaching, both in the content of the programme and in its pedagogy. Similarly, Setati proposes an alternative pedagogy. However, while this proposal operates at the level of social practice, her argument also focuses strongly at the level of social structure. The international and national status of the English language and the South African Language in Education Policy are presented as central to understanding teachers' practices and student and parent preferences, as well as to the design and development of the alternative pedagogy itself. Setati demonstrates that decisions about which language to use in the mathematics classroom are political (i.e. intervening at the level of social structure) as well as pedagogic (i.e. intervening at the level of social practice). As is often the case, looking at extreme situations highlights issues in ways that can inform our understanding more generally. In South Africa, the political role of language is highly visible, forcing our attention to the necessity of considering social structures, in order both to understand existing discursive practices and to develop more effective and equitable pedagogies.

In describing the chapters in this part, I have suggested that more substantial attention to analysis at the level of social structure would enhance our understanding of the reasons underpinning particular classroom practices and text-level characteristics, as well as our ability to address issues of equity. The relative lack of attention to social structure is not surprising, given that this is a characteristic of the field of mathematics education as a whole. Although the dominance of individual-level psychological theory has been challenged by the widespread adoption of sociocultural perspectives, mathematics education researchers within a sociocultural paradigm have tended to restrict their consideration of sociocultural factors to the level of local interactions or communities of practice taken in isolation from wider social structures. It is only relatively recently that research informed by social theory has begun to be accepted in major mainstream mathematics education conferences and journals (see Cotton and Gates 1996; Lerman et al. 2009). Moreover, forming and linking analyses at these different levels is challenging: what methodological tools are appropriate at each level and how are analyses at different levels to be related to one another? In what follows, I discuss some ways of thinking about discourse in mathematics classrooms that allow us to incorporate consideration of aspects of broader social structure as well as local classroom practices. In so doing, I hope to indicate how approaching discourse in this way highlights issues of equity and enables consideration of the key questions identified above: *Why do classroom interactions take the forms we observe and which students are advantaged or disadvantaged by such practices?*

1 Pedagogic Discourse and the Structuring of Classroom Discourse

There are, of course, various social theories that might form the basis of a structural-level analysis (see Chouliaraki and Fairclough 1999 for a review of these in the context of their development of critical discourse theory). When we are concerned with discourse in classrooms, however, the primary social practice of concern is that of teaching and learning, situated within broader social structures of local, national and international schooling systems. In analysing such discourse and identifying the issues for equity that it entails, it is necessary to take into account the particular nature of this type of social practice and its function in society as a significant site of socialisation and social reproduction. Here, I choose to draw on the construct *pedagogic discourse*, used in a specialised sense in work drawing on the theory of Basil Bernstein, who refers to "the rule which embeds a discourse of competence (skills of various kinds) into a discourse of social order" (Bernstein 1990, p. 183), forming a context in which social reproduction and production occurs. It is the principle by which mathematics is transformed into school mathematics, both selecting and shaping the mathematical content matter and 'embedding' it in the sets of rules, subject positions and relationships that structure classroom activity. In one sense, a "discourse without discourse" (Bernstein 2000, p. 32), pedagogic discourse does not have its own unique linguistic component, but nevertheless provides us with a means of describing and understanding the forces that shape classroom interactions. Pedagogic discourse may vary in the strength of the boundaries around the subject matter (what is defined as mathematics and how this relates to other areas of the school curriculum or practices outside the school) and in the strength and location of control over actions and interactions (for example, what kinds of behaviours are permitted for teachers and for students, who may decide which tasks to undertake and how to do them, who may initiate or evaluate activity). This construct of pedagogic discourse provides a tool for describing and distinguishing systematically between different forms of classroom practice, while considering how teachers and students may be positioned within such practices.

Variation in the strength of field boundaries (classification $C^{+/-}$) and of control of interactions (framing $F^{+/-}$) has been used to characterise 'typical' forms of pedagogy. Lerman and Tsatsaroni (1998) consider three such forms present in current thinking about mathematics education: *traditional* ($C^{+}F^{+}$), *progressive* ($C^{+}F^{-}$) and *radical* ($C^{-}F^{-}$). They argue that, despite drawing on child-centred educational philosophy (in the case of progressive pedagogy) or even recognising the validity of students' cultural knowledge (in the case of radical pedagogy), none of these forms avoids disadvantaging students from non-hegemonic social groups. The rhetoric surrounding some forms of progressive and radical pedagogic discourses with concern for equity often draws on notions such as agency, empowerment, student autonomy or teacher-as-facilitator.

Such notions challenge more traditional pedagogic discourses in which teachers maintain explicit authority over both subject matter and forms of interaction. However, because a lack of explicit teacher authority is often accompanied by implicit rules of interaction and implicit criteria for evaluating legitimate participation (in Bernstein's terms, *invisible pedagogy*), the valorisation of student agency *et cetera* can also act to obscure the inherent asymmetry in the pedagogic relationship between teacher and students. This is an asymmetry that may be realised both in the participants' relationships to knowledge – the teacher has the knowledge or at least has privileged access to the means of acquiring the knowledge[2] that is to be acquired by the students – and in the principles that structure classroom interactions – who may speak when, what kinds of things they may say and do, who may judge the value of what is said and done.

While recognising that offering students a greater degree of agency in the classroom may have benefits for their affective orientations to mathematics and to

[2] Even in forms of pedagogy that enable students to investigate with relative freedom, to determine the direction of their investigation and, hence, the specific mathematical knowledge they encounter, the teacher's level of expertise in the subject domain as a whole, as well as her institutional position, continue to give her a privileged position in relation to the 'new' knowledge brought to the classroom by the student. She is likely to have the competence to integrate such knowledge into her existing schema and hence to maintain her ability to evaluate the student's competence. She is also able, because of her institutional position, to rule whether this knowledge is to be considered relevant or legitimate within her classroom.

schooling, the major issue for equity is the question of the extent to which a particular form of pedagogic discourse positions students as potentially successful or unsuccessful learners and, hence, prepares them differentially for participation in society. In saying this, I am not envisaging 'society' as a static entity with a predetermined set of roles to be filled by the products of the education system. Participation in society may also involve critical and transformative action, challenging current roles and assumptions. A central question for educators concerned with equity is: What *forms of pedagogy might prepare* which *students to participate in such ways?*

Bernstein argues that invisible pedagogies are likely to disadvantage students from those social groups whose everyday patterns of language use are more distant from those commonly used in the school. Without explicit rules and criteria, such students are likely to lack access to the recognition and realisation rules that would allow them to distinguish and produce legitimate texts. Although invisible progressive pedagogies are often seen as the only alternative to oppressive traditional pedagogies, there are other possibilities that may be of interest for those concerned with equity – in particular, the notion of *radical visible pedagogy*. Bourne (2003) illustrates such a pedagogy in an English language class in an urban, socially disadvantaged context, showing how it combined markedly asymmetric teacher-led transmission, focusing explicitly on the principles of the specialised discourse to be acquired, with strong signalling of the shift between parts of the lesson with this specialised focus and parts when students were empowered to draw on their personal and collective experience to reflect and critique the specialised discourse. Bourne argues that such a form of pedagogy shifts the object of teaching and learning from individual attainment to collective endeavour. It thus has the potential to move away from reproduction of the status quo of social positions to construction of change between social groups.

The examples offered in the chapters in this part demonstrate asymmetry in different forms and in different aspects of the pedagogies. The glimpses we have of the classrooms studied by Setati and Barwell suggest fundamentally traditional pedagogies with strong framing, in which the teacher defines the task, the means by which students may legitimately respond to the task and the criteria for evaluating their responses. However, they differ in the extent to which an element of choice is devolved to students in producing their responses: whereas Barwell's 'K' has little room to manoeuvre as he attempts to participate in the teacher-controlled pattern of interaction, the students in Setati's classroom may not only choose the language they use to read and speak, but also have some opportunities to negotiate their means of engaging with the task as they work in groups independently of the teacher. In contrast, the pedagogy of professional development described in Herbel-Eisenmann's chapter is very weakly framed, allowing the teacher-students some choice in the particular focus of the subject matter as well as a large degree of control over the direction of the conduct of the development activity. Significantly, the reflexive discussion around 'telling' shows the author and the teacher-students recognising and struggling with asymmetry, both in their classrooms and in the professional development setting. It is this reflexivity that opens up the possibility for the participants to challenge and transform pedagogic relationships.

It is important to recognise that asymmetry does not necessarily entail inequity. Indeed, it could be argued that the reproductive and productive function of pedagogy is only possible in the presence of asymmetry in relation to knowledge. Of course, this raises the question of *whose* knowledge is to be reproduced and produced and here issues related to the identity and power dimensions of equity emerge. Radical pedagogic discourses open up the content matter of mathematics to recognise alternative, often local, forms of knowledge. Some emancipatory pedagogies, such as those based on Freirean principles and, in the field of mathematics, some varieties of ethnomathematics (e.g. Knijnik 2002), select and recontextualise mathematical knowledge according to principles rooted in the interests of the students and their local communities rather than in those of curriculum developers, academics, employers or governments. Alternatively, some critical mathematics pedagogies, again drawing on principles rooted in the interests of students and their communities, bring 'traditional' high-status mathematical knowledge into interaction with knowledge about society (e.g. Gutstein 2006; Skovsmose 1994). However, as Setati (2005b, this volume) points out, different forms of knowledge are associated with different forms of power. It is thus always relevant to ask whether and how a form of pedagogy allows students, in particular those from non-hegemonic social groups, to acquire powerful knowledge – and which kinds of power this may afford.

In the classrooms studied by Setati, the tension between access to political and socioeconomic power (through knowledge of the English language) and to epistemological power (through mathematical knowledge gained through the medium of familiar languages) is presented as an issue of choice between languages.[3] The pedagogy Setati proposes enables students to make that choice between languages for themselves, responding flexibly to their immediate need to make sense of their current task while recognising the generalised power of using the high-status English language. Students are thus enabled to negotiate their own paths towards more powerful positions. Nevertheless, the choices students make will still serve to differentiate those who become able to produce legitimate texts in both domains (legitimate both mathematically and linguistically) from those who, for example, engage mathematically through the medium of their home language but consequently do not develop their competence in English and hence are unable to achieve political power. In valuing student choice as a means of enabling access to highly valued forms of knowledge, it is nevertheless pertinent to ask both how the various choices available

[3] It is worth noting that this dilemma is not confined to contexts such as that of South Africa where the political power of language is obvious, due to its historical role as a site of struggle among the multiple communities in that country. In the United States, for example, debates about the status of African American Language face similar tensions among identity, epistemological access and power (see, for example, DeBose 2007), while programmes providing bilingual education for Spanish speakers are currently under legal and legislative threat.

to students may result in different trajectories and how such choices and trajectories are distributed among different groups of students.

Most often, of course, very little choice is offered to students about the nature of their learning opportunities. Decisions are made by curriculum designers, textbook authors, schools and teachers that shape and limit the forms of specialised mathematical discourse available to be acquired by various groups of students. This is most obvious where students of different kinds (defined, for example, by social class, gender, race or perceived 'ability') are structurally separated within the schooling system, whether physically separated in different schools or tracks or symbolically separated by the provision of differentiated resources. Dowling's (1998) analysis of a differentiated textbook scheme demonstrates starkly how the trajectories of 'high-ability' and 'low-ability' students are produced: 'high-ability' students being apprenticed to the specialised discourse of academic mathematics while their 'low-ability' counterparts are excluded from this, projecting future life paths in working-class occupations. Similarly, my analysis of extracts from textbooks designed for 'higher' and 'intermediate' groups of students (Morgan 2005) shows how the nature of mathematics and mathematical activity are construed very differently in the treatment of definition of trigonometric functions, offering 'higher' students access to identities as enquiring and potentially creative mathematicians, while 'intermediate' students may only receive knowledge from anonymous authority.

Studies of classrooms have also shown differences in the extent to which teacher–student interactions make use of specialised forms of mathematical discourse in schools with different social class and gender compositions, identifying more use of specialised mathematical forms in classes composed of students of higher social class status and male students (Atweh et al. 1998; O'Halloran 2004). I would not wish to claim that textbook authors or teachers deliberately seek to exclude working-class students, girls or other disadvantaged groups from learning high-status forms of mathematical discourse. Indeed, it seems far more likely that employing everyday discourses (not only in the form of everyday contexts for mathematical problems but also, more fundamentally, using an everyday register rather than the specialised lexico-grammar and forms of reasoning of the specialised mathematics register) is adopted as a strategy designed to enable such students to engage. The question that must be asked, however, is: *What is it they are to engage in and how will this affect their futures?*

2 Multilingualism and Classroom Discourse

Several of the chapters in this volume consider mathematics classrooms where the main language of learning and teaching is not the first language of some or all of the students. It is thus relevant to consider what specific issues arise in such settings. While I have argued that all pedagogic relations are inherently asymmetric, this is put into even clearer focus when we consider students attempting to learn mathematics in such classrooms. In multilingual contexts, there is additional asymmetry that arises from differential relationships to the language or languages of teaching and learning (who has language knowledge and who may decide which language to use in what circumstances?) and, as Setati powerfully points out, the languages themselves have different social, economic and political power beyond the classroom. Questions about what language is used in the classroom are thus not simply a matter of how effective communication may be achieved, but crucially also impact on the forms of knowledge that are valued and the projection of students' trajectories in the world outside the school.

As I have argued elsewhere (Morgan 2007), the designation 'multilingual' can be taken to have much wider applicability when we are interested in students' access to high-status forms of language and to mathematical discourse. Whether or not one wishes to accept the notion that learning mathematics is like learning a foreign language (cf. Ervynck 1992) or the rather more fully developed theoretical perspective proposed by Sfard (2008), equating thinking mathematically with engaging in mathematical discourse, it is nevertheless clear that achieving success in school mathematics necessarily involves learning to recognise, respond appropriately to and produce mathematical texts. It is thus a challenge for every learner to develop fluency in mathematical forms of discourse (or at least school mathematics discourse) and, more generally, in the forms of discourse that are valued in school. This challenge is substantially amplified when the learner does not have fluency in the 'national' language in which the mathematical discourse is embedded. It is also amplified for all those learners who lack fluency in the variety of language that dominates school discourses, whether or not their 'home' language is notionally the same. This includes speakers of 'non-standard'[4] varieties, such as African American Language, which have distinctive lexico-grammatical features. It also arguably includes children from working-class backgrounds whose home language practices are different from those valued in middle-class families and in the school.

The issue of multilingualism in mathematics education is usually posed in a way that assumes that, while languages may be diverse, there is only one mathematics. Indeed it has even been claimed that, whereas translation of other kinds of texts from one language to another is widely recognised to transform the meaning as well as the language, mathematical texts can be translated unproblematically (Layzer 1989, p. 126). Although this may be at least partially true if we restrict ourselves to considering academic pure mathematics texts dominated by formal notation, translated into 'world' languages (such as most European languages, Mandarin, Arabic, Japanese, etc.) with highly developed mathematics registers, it is certainly not the

[4] The term 'non-standard' clearly privileges high-status forms of language. In English-speaking countries, 'standard English', as spoken by the hegemonic group (at least in formal situations) and promulgated in national media and education systems, is taken to be 'correct' and desirable while other varieties, spoken by less-privileged groups, are perceived as lacking. Linguists have long established that non-standard varieties have their own coherent lexico-grammatical systems, yet their speakers are still accused of being careless or ignorant. The labelling of some varieties of language as non-standard is an exercise of power by dominant social groups.

case when we consider school-level texts that include a much higher proportion of 'natural' language and presentation of mathematics in everyday contexts – or when we consider a wider range of languages.

Halliday's original characterisation of the mathematics register (Halliday 1975a) was presented in the context of a UNESCO symposium focusing on the question of how to teach mathematics in (especially post-colonial) countries in which the national languages did not have such developed specialised mathematics registers. While attempts to develop such registers have often concentrated on inventing, adapting or adopting vocabulary to name mathematical concepts, participating in mathematical discourses involves much more than simply naming things. For example, Barton's (2008) exploration of the mathematical implications of linguistic structures in various non-European languages demonstrates that different lexico-grammatical characteristics can offer different potentials for construing meanings in relation to mathematically significant domains such as numerosity, temporality, spatial orientation and measurement (see also Wagner and Lunney Borden, this volume). Students whose first language affords meanings in such domains that are not consistent with those of hegemonic mathematical discourse may not only have to struggle with learning mathematics in a second or additional language, but also face the possibility that their mathematical ways of thinking are not compatible with those expected by their teachers and other gatekeepers.

This is not only an issue for students who are officially recognised as multilingual, but also for some of those who speak non-standard versions of a dominant language. For example, Orr's (1987) analysis of the work of her African American students identified them as failing in school mathematics, locating the blame for their failure in their 'deficient' variety of English (characterised in Orr's chosen title *Twice as Less*). Orr's interpretation has been strongly criticised for relying on a view of African American Language that is not consistent with current scholarship in linguistics (Baugh 1994). Nevertheless, most mathematics teachers are not experts in linguistics and deficit models of the languages of non-hegemonic groups are widespread in non-expert discourses.

However, it is not just lexico-grammatical differences that may be significant for speakers of different languages or varieties of language.[5] Indeed, in the context of debates about African American Language, Spears (2007) argues that, "the principal differences between Black ways of speaking and other forms of American speech lie in communicative practices" (p. 100). Spears notes in particular the creative freedom or 'semantic license' of AAL speakers: inventing new words or phrases or using old forms in new ways. At the level of genre, Cazden (1988/2001) followed up the work of Sarah Michaels, who identified differences in the typical structure of narrative in stories told by American children of European or African heritage. Significantly, Cazden reports that while African American adults recognised and valued both topic-focused and episodic narratives, their white counterparts

[5] I do not wish to get involved in debates about what constitutes a separate language, a language variety or a dialect.

responded negatively to episodic narrative (typically produced by African American children), suggesting that the child-author might have "language problems" and low academic achievement.

The challenge for multilingual learners and, indeed, for all those from non-dominant social groups with their own communicative practices (and, more broadly, their own cultural practices) is thus not just a matter of knowing or learning the national or international language, but also learning to make use of this language to communicate in the kinds of ways that are valued by teachers and by others with evaluative and gate-keeping roles. Of course, the way I have posed this as a challenge *for the learners* is an example of the behaviour of what Martin (2010) calls "white institutional space", accepting the hegemony of the forms of language, forms of mathematics and other values of the dominant group (white, Eurocentric, English-speaking, male, middle class) and assuming that members of other groups must learn to conform to them. Alternatively, this might be posed as a challenge for teachers and curriculum designers to learn to recognise and value the mathematical activity of those with a range of linguistic and cultural heritages. Such changes in the value system of school mathematics, however, would not just affect relationships within the practices of classrooms and schools, but would also threaten existing social structures – and may therefore anticipate concerted resistance from those with a vested interest in the current system. In an example of such resistance, Margaret Thatcher, then Prime Minister of the United Kingdom, was sufficiently concerned about challenges to the nature of the mathematics curriculum to condemn them in her 1987 speech to the Conservative Party Conference: "children who need to count and multiply are being taught anti-racist mathematics, whatever that may be".[6]

3 Finally

The proposition that is the title of this chapter, 'Studying discourse implies studying equity', is only true if you adopt a definition of discourse and an approach to studying it that, as I have insisted, embeds use of language within social practices and structures. With Bourdieu, I find it useful to think of language as part of cultural capital and, hence, to reject any approach "which reduces social exchanges to phenomena of communication and ignores the brutal fact of universal reducibility to economics" (Bourdieu 1986, p. 253). The ability to use powerful forms of language is thus associated with access to socioeconomic power and choices about which language or languages to use in the classroom may be seen to be political choices.

Of course, language also functions powerfully at a micro level in interactions between individuals and groups of individuals. Studies focused at this level in mathematics classrooms can provide us with important insights into the values of

[6]http://www.margaretthatcher.org/speeches/displaydocument.asp?docid=106941, accessed 13 July 2010.

the mathematical practices offered to students, into the nature of the pedagogic discourse and into the positions and identities available to students and teachers. Such insights may help us to understand how differences between mathematics learners are constructed locally through participation in particular forms of mathematical and linguistic practice and through evaluation practices.[7]

Remaining at this local level, however, reduces the descriptive, explanatory and predictive power of the analysis. The challenge is to connect such classroom-level analyses to a developed understanding of the broader context. By locating the analysis of local phenomena within a macro-level analysis of relevant social structures, it becomes possible to see how hegemonic discourses and the interests of dominant groups shape the pedagogic discourse. This more fundamental insight is necessary if we are to understand the forces with which we are likely to be confronted as we attempt to transform inequitable practices in mathematics classrooms.

[7]See, for example, Morgan and Watson (2002) in which Anne Watson and I discuss the consequences for equity raised by two studies of teachers' assessment practices.

Part III
Implications and Policy

Chapter 12
Equity, Mathematics Reform and Policy: The Dilemma of 'Opportunity to Learn'

Donna M. Harris and Celia Rousseau Anderson

In the United States, there have been substantial efforts to reform mathematics education, in part to address perceived shortcomings in international comparisons of mathematics achievement and in part to address longstanding educational inequities. A key theme in U.S. mathematics reform is to improve access to valued mathematical practices for groups that have been historically marginalized, according to their race, socioeconomic status or English-language proficiency. According to the U.S. National Council of Teachers of Mathematics *Standards* documents (NCTM 1989, 1991, 2000), these practices involve participating in the classroom discourse, including justifying and evaluating solutions to challenging problems. Certainly, teacher practices are critical for the ways discourse is structured in classrooms. However, in addition to teacher-level influences, we argue that access to these practices is shaped by structural factors beyond the level of the classroom.

When implemented equitably, discourse practices provide opportunities for all students to achieve mathematically. Research by Borman (2005), Boaler (2006b) and Boaler and Staples (2008) shows that mathematics instruction with discourse practices that promote more active intellectual roles for students leads to increased achievement and a narrowing of the achievement gap. Therefore, active student participation in classroom discourse as a central practice to changing teaching and learning in public schools has great potential to address the dual needs to improve mathematics education and expand access to high-quality mathematics for all. We start with the premise that implementation of these discourse practices by teachers

D.M. Harris (✉)
Department of Educational Leadership, Warner Graduate School of Education and Human Development, University of Rochester, Rochester, NY, USA
e-mail: dharris@warner.rochester.ca

C.R. Anderson
Department of Instruction and Curriculum Leadership, University of Memphis, Memphis, TN, USA
e-mail: croussea@memphis.edu

B. Herbel-Eisenmann et al. (eds.), *Equity in Discourse for Mathematics Education: Theories, Practices, and Policies*, Mathematics Education Library 55,
DOI 10.1007/978-94-007-2813-4_12, © Springer Science+Business Media B.V. 2012

who have an understanding of mathematics *and* issues that pertain to equity can lead to improved learning for all students. This chapter focuses on existing policy contexts that have the potential to limit student access to discourse and, as a result, impede the development of equitable learning opportunities.

This attention to policy-level concerns in mathematics education is tied to a vision of equity that focuses on opportunity to learn (OTL). In this chapter, we trace important factors that shape equity in mathematics education, especially in relation to discourse practices. These factors include tracking, the distribution of high-quality/qualified mathematics teachers and assessment policy. Attending to OTL factors is crucial to critically examining the structural conditions that affect access to valued classroom discourse practices. Through consideration of various OTL variables, we seek to make the case that systemic implementation of mathematics reforms, such as the recommendations to use reform-based discourse practices, will not contribute to equity until there is explicit attention to those structural conditions that continue to create disparate learning experiences and outcomes.

1 Equity and Opportunity to Learn

Through the OTL lens, we focus on access issues and the policy conditions that may either support or inhibit the implementation of discourse practices that promote more active and substantial intellectual roles for students. OTL is a conceptual framework used by many researchers (see Stevens 1993), though we build from the work of Pullin and Haertel (2008) to describe how we formulate OTL in this chapter. They describe three different conceptions of OTL. One is as a measure of the relationship between content taught and content tested: OTL, according to this view, is primarily a reflection of the students' opportunities to learn the content on which they are being assessed.

A second conception of OTL involves resources for educating: "OTL can also be defined in terms of *resources* to support teaching and learning, including teacher qualifications, technology, supportive services, expenditures and the like" (2008, p. 20; *italics in original*). They note that this attention to resources emerged in the 1980s and 1990s, as some researchers and policymakers questioned whether students should be held accountable for test results if their schools did not have access to the resources necessary to support the implementation of state curriculum frameworks adequately. The final conception focuses on instruction: "OTL can be defined in terms of classroom processes and practices enabling learning for individual students or groups of students" (p. 23). This is intended to provide a more expansive view of the nature of schooling and instruction. Pullin and Haertel state:

> a growing number of authors assert that meaningful OTL must be based on a theory of learning and models of teaching and schooling including instructional leadership and professional development, that take into account a full and rich understanding of the process and practices of education to attain meaningful results. (p. 23)

In our examination of OTL as related to discourse and policy, our focus is on the second and third (i.e. resources and classroom practices) conceptions of OTL. In particular, we consider both the resources that are available and the processes and practices that are necessary to improve the quality of teaching and learning. Resources such as teacher credentials and experience and money available are elements of OTL that are essential for improving the quality of teaching and learning. We suggest that low teacher capacity as reflected by credentials and experience will make the implementation of discourse practices difficult.

The view of OTL that considers instructional practices is reflected in Gee's (2008) attention to students' opportunities to acquire academic forms of language (also a view discussed by Barwell, this volume, and Schleppegrell 2004). According to Gee, OTL involves "learning not just the same 'content' but also [offering] equal affordances for action, participation and learning" (p. 104). Therefore, we must understand the interplay between students and their learning contexts to determine whether students can assimilate the content being taught. Similarly, Greeno and Gresalfi (2008) described OTL as "affordances for student participation that support trajectories toward stronger valued capabilities and dispositions" (p. 193). In this view, OTL extends beyond resources to include how those resources translate into learning opportunities in the classroom. Greeno and Gresalfi's conceptualization considers student interactions both in "communities of practice" (p. 170) and in the "material systems" (p. 175) students encounter within classrooms. These 'material systems' can be classroom activities where students engage in learning tasks to demonstrate knowledge.

Greeno and Gresalfi (2008) make several assumptions regarding OTL, including the notions that student learning evolves over time, builds on prior knowledge and stems from their interactions with the curriculum and other components of their material systems. Thus, all students are positioned as possessing pre-existing mathematics knowledge from everyday experiences to build on concepts introduced in classroom settings. They further suggest that learning is affected by learning-task rigor and student agency. For instance, tasks with low academic rigor will not build students' higher-order mathematics skills. If students are only expected to comprehend mathematics procedures, then such learning environments will more than likely produce students who will not have the capacity or willingness to engage in discourse activities that require them to be responsible for independent mathematics inquiry with their peers. This conceptualization of OTL (see also Cobb and Hodge 2002) supports a multi-layered view of equity, seen in terms of whether students have access to opportunities to develop forms of reasoning that have power (Bruner 1986).

These conceptions of OTL suggest that it is shaped by many factors and this complexity points to the need to examine OTL from various perspectives. For example, one consideration with respect to discourse and OTL is time. The process of reforming classroom discourse practices will be a lengthy one (Greeno and Gresalfi 2008; Nathan and Knuth 2003). Similarly, opportunities for teachers to learn how to implement discourse practices effectively are necessary. Nathan and Knuth's case study shows that adequate teacher support to change practice must be available for effective instruction. These are but two of the factors highlighted in an examination of OTL through the lens of resources and classroom practices. However, we argue that even these factors are shaped by larger structural conditions that we must consider. For, if sufficient conditions and resources are not available, then changed discourse practices as a reform to expand OTL will be severely limited.

2 Opportunity to Learn and Discourse: Implications of Tracking, Teacher Quality and Assessment Policies

In this section, we examine some of the policy-level factors that affect OTL with respect to mathematics reform more generally and discourse more specifically. In our examination of these factors, we consider policy (and practices) at a state or district level (e.g. assessment policies), as well as school-level (practices) policies (e.g. tracking). Our goal is to illustrate the role of policy in supporting or constraining OTL as it relates to discourse. For example, a study of teachers in high-poverty schools (Kitchen 2003) revealed several institutional conditions that operated as barriers to mathematics reform. Teachers reported that overwhelming workloads constrained the implementation of reform in their schools. Teachers in such schools lacked the time to plan and implement changes in their classroom practices. In addition, Kitchen also reported that the teachers were unable to get funding from administrators to attend meetings or other professional development activities. As we consider the relationship between equity and discourse, we must attend to the factors that may prevent individual or groups of students from obtaining access to high-quality instructional practices.

2.1 Tracking

One policy-level factor that appears to have an impact on students' OTL with respect to discourse is tracking. Tracking is commonly used to organize learning groups in K–12 schools to address the pre-existing academic differences among students, according to characterizations of high, average or low achievers. In high schools, tracking is often accomplished through more subtle curriculum differentiation. For example, there has been a push for all students to take Algebra. Some schools create multiple Algebra sections that have different designations, such as honors Algebra sections for high-achieving students. The concern about tracking is that students across learning groups are provided content that varies in academic rigor (Gamoran 1987; Oakes et al. 1992). For example, Oakes (1995) found that students in high-track mathematics classes had access to 'high-status' knowledge (ideas and concepts), whereas students in the low-track classes repeated the same basic computational skills year after year. Similarly, in a survey of teachers, Weiss (1994) discovered that low-track classes were significantly more likely than high-track classes to spend time each week doing worksheet problems and less likely t1o be asked to write or explain their reasoning about solving a mathematics problem. These differences between tracks were summed up by Oakes (1995) as she described the results from her study of two districts:

> Students in lower-track classes had fewer learning opportunities. Teachers expected less of them and gave them less exposure to curriculum and instruction in essential knowledge and skills. Lower-track classes also provided [...] students with less access to a whole range of resources and opportunities. (p. 687)

Thus, the existence of tracking limits the opportunity to learn for any student placed in lower-level mathematics groups. It is likely that the discourse patterns in lower-track classrooms will not be as engaging or challenging as those in higher-level classrooms, with most of the interactions being directed by the teacher and focused on mathematical procedures. Low-track mathematics classrooms are less likely to be contexts where student interactions are nurtured to foster agency for mathematics learning, limiting the opportunity to learn for lower-grouped students (see Hand 2010).

Since tracking tends to segregate students of color and low-income students into the lowest tracks, they are more likely to experience these inequities in opportunity-to-learn in mathematics classrooms (Braddock and Dawkins 1993). For example, Oakes (1995) demonstrated that the placement processes in two districts were racially skewed. African American and Latino students were much less likely than White or Asian students with the same test scores to be placed in high-track classes. She concluded from these findings that "grouping practices have created a cycle of restricted opportunities and diminished outcomes and exacerbated the differences between African American and Latino and White students" (p. 689). As a result of these practices, students of color within racially mixed schools continue to be disproportionately overrepresented in low-track mathematics courses (Oakes et al. 2000). If students of color and low-income students are in classrooms using less rigorous curriculum and instructional strategies, then there is little likelihood of closing the mathematics achievement gaps. Boaler and Staples (2008) found that achievement gaps *can* be closed when students are heterogeneously grouped and are offered challenging forms of curriculum and instruction, including interactive forms of discourse.

There is also evidence that tracking policies can impact not only which students become members of a particular class, but also how those students participate in the learning practices of the mathematics classroom. For example, Civil and Planas (2004) found that students' participation in the mathematics classroom was influenced by larger social and organizational structures. These structures, including gifted programs and special education, set up differential power and status positions within the classroom:

> The value of what is said and of who says it is established according to the place where each participant is located. Students respond according to where they are supposed to belong. Students placed in the high-status system may have easier access to the mathematical discourse, whereas students placed in the low-status system are still supposed to prove their value. Roles influence the students so that they learn to act and behave in ways that agree with the social order of the educational arrangements. (p. 11)

This suggests that the stratification of students into learning groups influences students' self-perceptions regarding their capacity to do mathematics. The social stigma associated with low-level track placement may limit student willingness to participate in mathematics classes using discourse practices. Chazan (2000), for example, found that his school's organizational structures shaped students' participation in his low-track high school mathematics class. In particular, he discovered that the students did not easily engage in discourse practices. They were reluctant to share their thinking publicly and were not in the habit of listening to the thinking of other students. Chazan viewed both of these dynamics as related to the notion of ability that had been constructed through the tracking practices of the school. He hypothesized that students who had previously been judged by teachers to be of low ability and, consequently, placed in low-track classes were reluctant to share their thinking, perhaps in fear of the same types of judgments. His students also questioned the purpose of listening to other students, like themselves, who had been judged to be of low ability. Thus, the tracking policies of schools can influence OTL in the form of the participation patterns within the classroom, whether by creating a status structure within more heterogeneous classrooms or by constraining overall interaction in low-track classes. Moreover, Civil and Planas (2004) pointed out that the students most negatively affected by the social order created through institutional grouping practices are members of particular ethnic or language groups or those of lower socioeconomic status.

Proponents of tracking (Hallinan 1994) suggest that tracking allows teachers to meet the needs of students effectively. There is a concern that high achieving students lose out when heterogeneously grouped. However, Boaler (2003, 2006a, b) and Boaler and Staples (2008) indicate that heterogeneously grouped students in California classrooms implementing discourse practices and a problem-based curriculum to guide instruction were more enthusiastic about mathematics and showed significant gains in outcomes compared with students in tracked mathematics classrooms using traditional curriculum and instruction. Moreover, Boaler and Staples found that the highest achieving students in classrooms using more interactive and problem-centered forms of curriculum and instruction made the greatest gains in mathematics outcomes in the second year of their study compared with students in traditionally tracked classrooms. In reform-based contexts, the mathematics achievement gaps among White, Latino and African American students disappeared, though Asian Americans continued to achieve higher than other groups. However, in classrooms using traditional mathematics instruction with tracked mathematics curriculum the racial achievement gaps persisted. This evidence suggests that tracked mathematics classrooms using a traditional mathematics instructional focus limit students' *relative* opportunity to learn.

2.2 Teacher Quality

Another OTL factor that is likely related to the effective implementation of discourse practices is teacher quality. Teachers are an essential resource related to opportunity-to-learn. When we consider the mathematical and pedagogical demands of orchestrating classroom discourse and other aspects of teaching for understanding, the issue of teacher knowledge and teacher quality cannot be overlooked (Ball et al. 2008). As Ball and her colleagues note, teachers use a wide range of knowledge on a daily basis in the act of teaching mathematics for understanding.

Our concern from a policy standpoint relates to the inequitable distribution of teacher knowledge that potentially impacts students' access to high-quality discourse practices. For example, African American and Latino students are more likely to be taught by teachers without a degree or certification in mathematics (Darling-Hammond 1995, 1997; Lee 2004). Moreover, while out-of-field teachers instruct over one-fourth of high school mathematics students, these proportions are highest in high-poverty schools and high-minority classes (Darling-Hammond 1997). In a national survey, Weiss (1994) found that more than half of high-school mathematics classes with populations that were at least 40% minority were taught by teachers without a degree in the field. Ingersoll (1999) found similar results in high-poverty schools (schools in which 50% or more of the students receive free or reduced-price lunch). Whereas 27% of secondary mathematics teachers in low-poverty schools (schools in which 10% or fewer of the students receive free or reduced-price lunch) had neither a major nor a minor in the field, approximately 43% of teachers in high-poverty schools did not have at least the equivalent of a minor in the field. Teacher inequality is also related to tracking and achievement levels. In particular, low-track and low-achievement classes are more likely to have more out-of-field teachers than high-track and high-achievement classes (Ingersoll and Gruber 1996; Oakes et al. 2000).

We suggest that under-qualified teachers are less likely to orchestrate a classroom that has the participation systems and task rigor to anchor high-status discourse practices. Teachers' capacity to implement discourse is not evenly distributed among school personnel despite professional development efforts explicitly aimed at helping teachers' discourse practices. Khisty and Chval's (2002) examination of two teachers' classroom discourse practices provides insights into how a veteran teacher was able to effectively help students develop mathematics language, while an inexperienced teacher was less successful in helping her students develop mathematics language learning, even though both teachers received professional development regarding discourse practices. Although the inexperienced teacher had set up some of the processes needed to guide instruction using mathematics discourse strategies, she dominated classroom discussion and failed to develop appropriate mathematics language in her classroom. This study illustrates how less experienced and under-qualified mathematics teachers may have difficulties providing students opportunities to learn mathematics through engaging discourse practices.

We acknowledge that measures such as certification status and degree in the field are only proxies for the knowledge and depth of understanding necessary to teach effectively. However, the predictive power of these indicators with respect to student achievement and the disparities in their distribution highlight the significance of teacher quality to any discussion of OTL in mathematics education. We should be clear, however, that discussions of teacher quality are not intended to put the blame for low student achievement or lack of mathematics reform on teachers themselves. Rather, decisions regarding emergency licensing and the distribution of less experienced teachers are made at a policy level by district and state administrators. Moreover, the policies that contribute to teacher shortages and other conditions influencing teacher hiring are most often made at district or state levels and point to more systemic inequalities in school funding and resource allocation that disproportionately impact students of color. Our point is not to blame the teachers, but rather to note the influence of these policy-level decisions on the conditions influencing OTL.

2.3 Assessment Policies

OTL is also influenced by the assessment policies of the school district or state and sanctions that respond to the U.S. federal policy demands of No Child Left Behind (NCLB). The stakes associated with NCLB are supposed to be the incentive for improved teaching and learning that states impose on local schools and districts (Fuhrman 2004). However, the consequences associated with assessment policies may hinder the development of discourse practices in ways that level the playing field. In Texas, for example, Haney (2000) and McNeil and Valenzuela (2000) reported that teachers in their studies spent a substantial amount of time preparing for the state test (the TAAS). Teachers in Haney's study indicated that they began test preparation more than a month before the test. According to McNeil and Valenzuela, teachers in their study reported spending several hours a week drilling students on practice exams. In this effort, commercial test-prep materials became the *de facto* curriculum in many schools, reducing mathematics to sets of isolated skills. Teachers reported that the time devoted to instructional activities that engage students in higher-order problem solving was severely reduced (or disappeared completely) in the press to prepare students for the TAAS.

From an OTL perspective, this narrowing of the curriculum in response to assessment is particularly problematic, as it is more likely to occur in high-minority classrooms (Darling-Hammond 1994; Lipman 2004; Madaus 1994; McNeil and Valenzuela 2000; Rousseau and Powell 2005; Shepard 1991; Weiss 1994). For example, in a study of two predominantly African American schools, Lipman (2004) found that teachers were under substantial pressure to improve performance on the mandated standardized test. One school was on 'probation' for previous years of low achievement on the test and the other was in danger of being put on probation. In both schools, the impact of the test was pervasive. Classroom instruction and curricular choices were driven by the test. Test preparation materials were substituted for the existing curriculum and practice in test-taking skills was a routine classroom activity.

According to Lipman, "achieving high test scores was equated with good teaching" (p. 81). In such situations, teachers often abandoned classroom practices, such as discourse practices that involve challenging intellectual engagement and that call on the students as resources, to meet the demands of high-stakes tests

(Rousseau and Powell 2005). Yet, Lipman notes that this pressure was not felt equally by all schools. Rather, the influence of testing on the culture of the school was more likely in low-income schools and/or schools populated by students of color. In contrast, relatively affluent schools were more likely to be immune to such pressures: when "probation schools and low-scoring schools serving low-income students of color are defined by test preparation, accountability policies widen the existing gap between them and schools serving middle-class and white student populations" (p. 85). The pressure to get schools off probation resulted in narrow instructional choices that reinforced traditional teaching methods rather than high-quality discourse practices.

3 In Conclusion

Discourse practices implemented across classrooms in the United State have great potential for improving mathematics education and reducing or eliminating disparities among students. However, opportunities to learn cannot be increased if there are structural conditions that limit the application of high-quality discourse practices. Tracking, the distribution of teacher quality and assessment policies can create school and classroom conditions that make the implementation of high-quality discourse practices unlikely. These policy contexts interact with the implementation of discourse practices in mathematics and contribute to continued stratification of learning opportunities. Since most schools in the United States use some form of student differentiation, we can expect that students labeled as 'of low ability' will continue to have limited access to high-quality discourse practices. Buckley's (2010) study shows how mathematics teacher expectations for students were mediated by tracking, despite efforts that the mathematics department made toward providing more equitable access to course offerings. Teacher expectations for low-tracked students remained, limiting access to academically rigorous content. Under these conditions it is unlikely that the implementation of high-quality discourse strategies would take place in low-track classrooms.

Since the distribution of teacher quality varies across school contexts and is associated with the racial and socioeconomic composition of schools, different investments must be made to ensure that mathematics teachers possess the content knowledge and pedagogical support to implement discourse practices. Teachers in low-performing schools with large concentrations of the poor and students of color must obtain additional support sustained over time when implementing discourse as a reform effort, since there are many barriers to overcome. Increased investment to develop and/or recruit well-qualified mathematics teachers to serve in low-performing school systems must be made in order to address inequities in teacher distribution.

Accountability pressures that result from assessment policy and the responses by districts, schools and teachers encountering probation status create learning contexts that are not ripe for the use of discourse practices. Assessment policies can contradict the equity aims of discourse and result in an incomplete view of opportunity to learn. In chronically low-performing schools, OTL is usually narrowly focused on those basic skills that are found on state assessments. Although students may be better prepared to pass the state assessment, the conditions needed to allow for all students to engage deeply with mathematics have not been addressed. As a result, a fundamental flaw of assessment policy is that many assume that the test pressure is enough to improve teaching and learning, and so less attention is given to the types instructional processes needed to drastically improve learning, where students are deeply engaged with mathematics.

In conclusion, we assert that these policies that operate to constrain equity and OTL with respect to discourse cannot be overlooked in the implementation of mathematics reform. Failure to consider these policies can inadvertently recreate the inequities that students have experienced within a more traditional system. If we do not attend to the systemic factors that influence the implementation of discourse, we may simply perpetuate inequitably distributed opportunities to learn.

Chapter 13
Educational Policy and Classroom Discourse Practices: Tensions and Possibilities

Jeffrey Choppin, David Wagner, and Beth Herbel-Eisenmann

> A central question for educators concerned with equity is: *What* forms of pedagogy might prepare *which* students to participate in [critical and transformative action]? (Morgan, this volume, p. 186)

What forms of discourse are valued? And by whom? Who gets to determine which forms of discourse are valued? Who gets to participate in the valued forms of discourse? And who controls participation? Can policy impact students' opportunities to participate in valued discourse practices? These questions focus attention on how students are positioned in classroom interactions, positioning that involves power relations that potentially develops or stymies the formation of students' mathematical identities and that ultimately has implications for students' achievement and access to resources within and beyond school mathematics. We take the term 'policy' to refer to a wide range of governmental or administrative efforts to mandate or otherwise influence particular practices in school mathematics classes.

With respect to these questions, in this chapter we frame a dialogue between researchers and policymakers. We consider tensions among researchers and policymakers, particularly ones that relate to the scope and level of attention and to the proximity to classroom interactions. Although policymakers are often concerned with mandating classroom practices at scale and from a distance, many mathematics

J. Choppin (✉)
Department of Teaching and Curriculum, Warner Graduate School of Education and Human Development, University of Rochester, Rochester, NY, USA
e-mail: jchoppin@warner.rochester.edu

D. Wagner
Faculty of Education, University of New Brunswick, Fredericton, NB, Canada
e-mail: dwagner@unb.edu

B. Herbel-Eisenmann
Department of Teacher Education, College of Education, Michigan State University, East Lansing, MI, USA
e-mail: bhe@msu.edu

B. Herbel-Eisenmann et al. (eds.), *Equity in Discourse for Mathematics Education: Theories, Practices, and Policies*, Mathematics Education Library 55,
DOI 10.1007/978-94-007-2813-4_13,

education researchers, including some featured in this volume, focus on interactions within particular classrooms. These differences have an impact on how equity and discourse are framed in the various communities, with implications for the scale at which change is envisioned or evaluated. Given the relative access to resources available to policymakers, however, these differences are not neutral with respect to students' opportunities to learn mathematics. Consequently, we explore the ways that researchers and policymakers might speak to each other, with the intent of highlighting possibilities for making policy more responsive to particular contexts and for making research more responsive to policy concerns.

We recognize that discourse and equity are inextricably intertwined. Moschkovich, in this volume, questioned the tendency to separate discourse and equity, pointing out that doing so often marginalized the discourse practices and research activities situated in the classrooms of liminal communities, as if discourse practices in these classrooms lacked legitimate mathematical forms of reasoning. She explains how the situated and hybridized nature of mathematical discourse practices binds together identity, power, access and achievement, which comprise Gutiérrez' (this volume) four dimensions of equity. We explicitly build from these dimensions of equity to provide a frame of reference for situating the tensions within and between these two communities.

Policy, for example, primarily addresses the tensions lying along what she called the *dominant* axis between access and achievement, especially with regard to addressing inequitable circumstances. Classroom-based researchers, on the other hand, have often focused on the tension between the dominant axis and what Gutiérrez called the *critical* axis, exploring the need to attend to identity and power (the two arms of the critical axis) as precursors to discussions of access and achievement. The nuances of such research findings, however, especially with regard to the role of context, seem to be rarely evident in policy.

As mathematics education researchers who have focused our work at the classroom level, we start with the view of 'equitable discourse practices' as being responsive to students' identities and intellectual resources, as well as to the social, cultural, historical and political contexts in which students operate. We thus acknowledge our agreement with Gutiérrez that equity cannot simply be framed in terms of access and achievement, but must also include issues of identity and power.

In this chapter, we explore characteristics of policy and research that undergird these tensions. We place these voices in dialogue to underscore both tensions and possibilities. One set of voices is associated with the policy community and includes policymakers as well as policy scholars whom we interviewed for this chapter. In the interviews, we posed a series of questions, such as: *Can policy alter classroom practices in ways that address long-standing inequities? Should policy focus on helping teachers gain a better understanding of and develop competency in new practices?* We identified policymakers from a range of perspectives and international contexts. Because of our selection of experts, our examples are drawn from four countries: Canada and the USA (neighbouring countries in which we and some of our policy experts work), and Bhutan and China (neighbouring countries in which two of our policy experts work).

We note that attention to equity and to discourse in mathematics classrooms differs across countries. For example, recent PISA results identify large disparities in student mathematical achievement in the U.S.A. that align with socioeconomic disparities (Education Trust 2010). By contrast, neighbouring Canada has relatively little disparity attributable to socioeconomic differences. Thus, there is little surprise to us that equity has become a critical focus for scholars and policymakers in the U.S.A.

The second set of voices primarily includes the authors of the chapters in this volume. These chapters raise important issues about discourse and equity in mathematics classrooms in a range of international contexts and we think it is important to connect their voices in dialogue with policymakers to consider what might *be done* to influence the nature of classroom discourse at a broader scale than what was reported in many of the individual chapters in this volume. We acknowledge Moschkovich's recommendations in Chap. 6 for how to think about and conduct research on mathematics discourse practices, yet make more explicit connections to the policy conversation at the heart of this one. Considering our developing sense of the centrality of discourse practices to equity and inequities, what might leaders in mathematics education do to affect the nature of classroom discourse practices? Although some chapters in this book provide examples of leaders working to support and/or structure discourse in particular contexts (e.g. Herbel-Eisenmann; Wagner and Lunney Borden; Setati), this chapter takes a broader look at the question.

1 Framing the Dialogue

Dialogue entails that actors are positioned with respect to external criteria and in relation to each other. We use Gutiérrez' dimensions of equity to frame criteria to situate each community and then we see how the communities relate to each other. First, we note that studies relating to the *dominant* axis, more so than studies relating to the critical axis, address dimensions of equity that are objects of measurement, and thus are more likely to be subject to large-scale interventions (e.g. policy) crafted at a distance from the contexts in which they are to be implemented. Conceptions of achievement often involve standardized measurements, such as standardized test scores, while conceptions of access usually involve variables such as course-taking and teacher quality (see Harris and Anderson, this volume) that can be measured with little contextual nuance.

Conversely, identity is often explored interactionally at small grain sizes, due to the complex and contested nature of both the term and the ways people construe themselves in the world. Also, it is enacted locally. Similarly, power is not easily measured or studied at scale, as it operates relationally (Foucault 1975/1984) in formal and informal interactions among actors situated within and between layers of organizations. Consequently, issues of scale and proximity are central to the ways policy and classroom-focused research are constructed, with implications for the perspectives of people operating in each community and for the ways these perspectives engage with the various dimensions of equity.

2 Tensions Between Access and Achievement

We begin the discussion by introducing our selected policy informants and then describe how some of them have addressed the impact of policy on students' opportunity to learn. We point out how these experts described the constraints and affordances of policy before turning to the perspectives outlined by the classroom-based researchers.

We chose to talk with policymakers and scholars from a broad array of policy perspectives and contexts, listed below:

- Doug Willms, Professor at the University of New Brunswick, Canada, who does large-scale quantitative analysis of education in relation to social policy, using Canadian and international databases of social demographics and of education achievement results and associated surveys;
- Karen King, Director of Research at the National Council of Teachers of Mathematics, U.S.A., who has worked as a Program Director in the National Science Foundation and served on the RAND Mathematics Study Panel[1];
- Karma Yeshey, Director in the Curriculum Office in Bhutan's Ministry of Education, who leads mathematics curriculum reform in Bhutan, a country undergoing major national mathematics curriculum reform;
- Lynn Paine, Professor at Michigan State University, U.S.A., who has studied the Chinese educational system and mathematics teacher inductions systems, as well as new teacher induction forms, processes and policies in many countries;
- Walter Secada, Professor at the University of Miami, U.S.A., who has studied equity in mathematics education, particularly around how policies impact English language learners.

Two primary mechanisms by which policy impacts equitable opportunities to learn mathematics were noted by the experts who addressed this issue. The first is the regulation of practice by means of creating incentives and penalties related to high-stakes standardized achievement measures, as is the case in the U.S. context. The second occurs through developing local capacity to enact the mandated curriculum, with an emphasis on supporting schools to help students learn challenging mathematics.

The distinction between these mechanisms is exemplified by the purposes for which educational authorities mandate curriculum. In China, the mandated curriculum has sometimes served as a focal point for developing teachers' capacity to understand relationships among curriculum, instruction and learning (Wang and Paine 2003). Paine, for example, noted in her interview that a primary goal of local educational policy in China is to provide resources that help teachers develop the

[1]This committee's charge was to make recommendations to the U.S. Department of Education about future research funding in mathematics education.

capacity to teach the mandated curriculum in ways that provide improved opportunities for students to learn mathematics:

> [The policy] was saying something is getting in the way of these kids learning and these teachers being able to support their learning, and let's figure out policies that might support the conditions that could or create the conditions that could support them [...] Now the goal is clearly equity, [...] concern that the larger portion of kids weren't achieving at the level that people might hope, but the problem wasn't the kids, and the problem wasn't even entirely the teachers, but it was a capacity issue where there needs to be learning, so there was a policy geared towards learning, which feels different from a policy geared towards measuring outcomes. What got produced was different [in many contexts], so in some cases it meant teachers were seconded from school A to school B to spend time teaching in a very different school [for a period of time ...] or a principal would be assigned to become the principal of two schools or to leave the school and work at another school and it worked out differently in different places in [Shanghai]. That seemed like a really novel approach [...] but the focus of the policy was very different from [the way we focus policy in the U.S.].

In her research, Paine describes how Chinese teachers form learning groups (similar to lesson study) around "learning how to think about an important topic and content and then try to understand what is hard for kids about this content, [which they do] by actually interviewing kids who have studied that topic in the past". They then use this information as data to improve the teaching of those topics. The overarching goal of these learning groups is to figure out how best to teach every child, particularly with respect to content areas in which many students struggle. Additionally, she explained how the teachers worked together to develop formative assessments that could be given at different points in time, drawing on this data to understand better what students were learning and using that information to inform their instruction. This kind of professional development, Paine argues, provides an interesting example of policy that supports issues of equity and discourse in mathematics education, but also points out that the process was labour intensive (see also Paine et al. 2003). We note that in China, relative to the U.S., the corrective to inequities involves providing schools and teachers with greater autonomy to improve instruction for all students. Policy in China includes a structure that mandates that the voices of students and of teachers are valued and instrumental.

In other jurisdictions, there are aspects of such teacher autonomy that co-exist with forms of top-down control. For example, in Canada there are clearly defined curriculum outcomes that describe mostly the mathematical procedures that students should be able to perform. The province of New Brunswick's curriculum states: "By the end of grade 5, students will be expected to divide 2-, 3- and 4-digit numbers by single-digit divisors and investigate division by 2-digit divisors" (NBDOE 2001, p. 30). This and the other outcomes are enforced in a way by external common assessments, though the results have less explicit implications on funding and programs than they do in the *No Child Left Behind* (NCLB) era in the U.S. Nevertheless, Canadian jurisdictions have legislation that guarantees teachers autonomy regarding ways of addressing mandated outcomes: "The duties of a teacher employed in a school include [...] identifying and implementing learning and evaluation strategies that foster a positive learning environment aimed at helping each pupil achieve prescribed learning outcomes" (PNB 1997, para. 27.1) There is a tension between external control and externally-mandated authorization of the teacher's professional responsibility to make decisions. A similar circumstance holds in France (see Pimm 2003), where the curriculum is nationally specified, but the means of instruction are not.

In other contexts, such as in the U.S., the mandated curriculum is intended to regulate – with little attempt to support – teachers' practices through the mechanism of standardized testing, which is particularly evident in contexts that serve marginalized students (see Harris and Anderson, this volume). Policy in the U.S. is insensitive to context, especially in terms of supporting the development of capacity in under-resourced contexts, which is particularly problematic since the U.S. has one the largest gaps in the world between the performance of low- and high-SES students (Education Trust 2010).

2.1 Affordances and Constraints of Policy with Respect to Access and Achievement

We now explore in more detail the affordances and constraints of policy in the U.S. context as an example of an environment that emphasizes external control. We explore, in particular, how policy attempts to address the inequitable distribution of access to high-quality curriculum and instruction, which we associate with an achievement gap that follows socioeconomic lines.

Even when the goal of policy is aimed at ameliorating inequitable circumstances, it is unclear exactly what policy may accomplish in this regard, especially if the policy does not articulate how schools and teachers can develop the capacity to enact challenging forms of curriculum and instruction for students. U.S. policy scholars Walter Secada and Karen King, for example, discussed some of the affordances and constraints of attempts in the U.S. to enact policies related to access and achievement. They both described policy as a 'blunt instrument', one that does not differentiate among local contexts or teachers.

Policies in the U.S. context, such as the *No Child Left Behind* legislation, mandate progress toward specific levels of achievement, which typically involve the use of standardized assessments and prescribed curriculum content to measure that achievement. The downside to such prescription without accompanying support mechanisms has been well-documented with regard to narrowing the kinds of curriculum and instruction to which students have access, especially in schools in which there are high percentages of minority students and high poverty rates (see Harris and Anderson, this volume). Yet, as Moschkovich highlights in her chapter, there are important alternative forms of assessment that might be better suited to contexts in which students are learning English in addition to mathematics. These alternatives are rarely considered in policy contexts.

Secada suggested that the 'bluntness' of policy is the result of attempting quickly and radically to change people's behaviour. However, the blunt nature of policy is endemic to the nature of educational systems in the U.S., in part due to how government is organized. In the U.S., for example, education governance structures are ultimately local, which leads to considerable variation between districts and schools in terms of how state and federal policies are interpreted (Cohen 1995). By contrast, jurisdictions are larger in Canada – they are provincial – and in Bhutan and China, the system is national.[2] As policy gets pushed closer to classrooms in the U.S., what gets transmitted is typically at best an awareness of reform doctrine, with little emphasis on how to support teachers in developing new practices around the doctrine (Cohen 1995). Consequently, there is likely to be little coherence *in practice* at the level of the classroom.

Spillane (1998) goes further, noting the segmentation *within* schools and districts that causes variation within those organizations with respect to policy implementation, making it less likely that there will be coherent messages or well-designed support systems for teachers as they enact new policy. Although local governance in the U.S. has the potential to be responsive to local contexts, so far we and our policy discussants see that it has not successfully addressed equity in a coherent and centralized way.

Also in the context of the U.S., King described policy as being unresponsive to conditions in local contexts. She pointed out, for example, that policies might be made to mandate aspects of classroom discourse as they relate to specific groups of children who historically have been marginalized by schools and that these policies could lead to "essentializing children, making all children of type X [the same], assuming there is some essential feature of them that they can build on". She argued that, instead, a more appropriate approach to classroom discourse should involve "having genuine conversations with children" and families in the community. King's points echo the recommendations made by Moschkovich in Chap. 6.

King similarly described the undifferentiating aspect of professional development in the U.S. that is not responsive to the local demands of teachers. She claimed that the collective set of teachers' professional development experiences are haphazard and unorganized, rather than building toward a common goal of expertise in an area of practice:

> [It is] not surprising that policy finds that people plateau in five to seven years because they get the same stuff they've always been getting. There's no catalyst to do something different. […] A policy regime that would help with differentiated professional development for different expertise would make more sense than the ones that are typically in place. Particularly, the ways in which we would expect beginning teachers to attend to discourse and equity would be different from more experienced teachers who have the freedom of mind to think about broader issues, like who is participating, the level of discourse happening. The system doesn't make it easy to help an experienced teacher to advance. They have to seek these experiences out.

King stated that although NCLB mandates the narrowing of the achievement gap, it ignores what happens in urban settings and consequently reduces students'

[2] The population of Bhutan is similar to that of a small province and smaller than many cities in the U.S. and Canada, but the connections between authorities and schools are more distant, both organizationally and geographically, due to challenging mountain terrain.

opportunity to learn by constraining the forms of curriculum and instruction. The lack of flexibility in the law reduces teachers' ability to react to the specific needs and resources of their students, removing the potential or even possibility for teachers to exercise professional judgment.

Secada, however, noted that such policy is sometimes a necessary evil, because it directs attention to important issues:

> And so policy, which is a blunt instrument – and it's a horrible instrument from the standpoint of forcing things to happen – is, in fact, an instrument that opens up spaces […] and makes possible asking those questions and having people say, 'Yes this is an important question'.

He described an example of educators needing to attend to the challenges experienced by mathematics students who are also learning to speak English, which has clear equity implications:

> Before *No Child Left Behind*, I had to beg people to come to anything I did about teaching math to English language learners. I would go to major urban districts that were under court order to desegregate along the lines of language and I would go there to do anything on teaching math to English language learners. The people in the math departments would tell me, 'That's the business of the bilingual people.' The people in the bilingual department and ESL department would tell me, 'That's the job of the math people.' If I was lucky I might get maybe 20 to 30 people… Now I do six sessions with 50 to 100 teachers in each session about how to do things involving the teaching of math to English language learners.

Attention to equity in the U.S. is also prevalent in research. Some of that attention may be attributable to gaps identified by policy instruments, such as the disparity we mentioned above. However, researchers who have close associations with teachers who face the disparities every day are responding to more than published results exposing inequities. They are responding to the realities they see among students and teachers.

2.2 Classroom-Based Researchers' Perspectives on Access and Achievement

In this sub-section, we discuss the perspectives of mathematics education researchers who explore equity and discourse in mathematics classrooms. These researchers critique the nature of policy with regard to access and achievement, for reasons that have both dissonance and resonance with those of policymakers.

The classroom-based researchers in this volume expressed a variety of perspectives about policy and how it gets interpreted, particularly with respect to the ways policy frames access to conventional forms of mathematical content and terminology. These perspectives lay bare the tensions within the dominant axis – namely, that access to particular forms of mathematics somehow leads to achievement – as well as the tensions between the two axes, explored more fully in the following section. In short, the perspectives expressed by mathematics education researchers concerning access point to the perils of ignoring political and cultural dimensions of policy (also discussed in more detail in the subsequent section), as well as the perils of ignoring access to dominant forms of mathematics.

In Chap. 9, Barwell criticizes the notion of 'access' as a basis for formulating policy, particularly around classroom discourse for students in the U.K. who are learning English. He states that policy that does not take into account the broader cultural and linguistic milieux in which students learn mathematics is narrow and potentially harmful. He states that policy needs to move away from deficit- or access-based notions of language with respect to mathematical learning. In order to understand fully the mathematics learning of learners of English as an Additional Language (the U.K. term), for example, Barwell states that policy-makers need to consider language demands besides vocabulary, to look at the demands of learning to use registers or genres (also see the chapters from Moschkovich and Schleppegrell) and, additionally, discursive demands.

Barwell describes *discursive demands* as "broader discursive aspects of bilingual, multilingual or second language mathematics classrooms, including the use of multiple languages; the role of students' everyday language; the interpretation of graphs, tables and diagrams; the construction of students' relationships with each other; and political tensions surrounding language use" (p. 151). Barwell criticizes policy in England as too focused on access, with potentially damaging impact on students for whom English is not their first language. In part, his disagreement arises because notions of access construct language use in simplistic terms and ignore the more complex and situated features of language use that have been shown to be useful, if not imperative, to the learning of mathematical concepts.

Setati, in Chap. 8, describes how teachers in South Africa interpreted policy that was ostensibly intended to incorporate students' home languages in multilingual contexts. Her research shows the limitations of policy that does not take into account the political aspects of access and achievement. She details the tensions for teachers between providing access to dominant forms of language and responding to policies that emphasized the use of multiple home languages.

Setati notes that, although the "South African Language in Education Policy (LiEP) recognises 11 official languages and encourages multilingualism, as well as language practices such as code-switching, as resources for learning and teaching in multilingual classrooms" (p. 128), there was little evidence of these practices in classrooms. The reason for the disjuncture between policy and practice, she explains, is that the policy assumes "that multilingual mathematics teachers, learners and parents are somehow free of economic, political and ideological constraints and pressures when they apparently freely opt for English as the LoLT [Language of Learning and Teaching]" (p. 131).

Setati observed the prevalence of English use in multilingual classrooms and cited the teachers' ideological and pragmatic reasons for preferring to teach in English, rather than in students' home languages. Consequently, the teachers' preferred language practices contravened what research indicates is helpful to the learning of mathematics. She explains:

> There seems to be a tension between the desire to gain access to English and the important, but not always recognised and acknowledged, need to gain access to mathematical knowledge. (p. 132)

Barwell, Setati and Moschkovich each emphasize the importance of using multiple languages and genres to learn mathematics: that is, although the ways policy was articulated or interpreted was intended to provide access to dominant forms of mathematics, these efforts had the unintended effect of restricting access to forms of language use that actually facilitate mathematics learning. However, there are risks to allowing learners to rely on non-dominant languages and genres to learn mathematics. Morgan, in her chapter for example, states that it is clear "that achieving success in school mathematics necessarily involves learning to recognise, respond appropriately to and produce mathematical texts" (p. 189). However, if students are allowed to make choices in how they produce mathematical texts, especially with regard to high-value languages and particular registers, there are consequences:

> the choices students make will still serve to differentiate those who become able to produce legitimate texts in both domains (legitimate both mathematically and linguistically) from those who, for example, engage mathematically through the medium of their home language but consequently do not develop their competence in English and hence are unable to achieve political power. (p. 187)

Morgan explains in Chap. 7 that these choices may result in different trajectories and differences in how those trajectories are distributed across groups of students, with potential consequences for who ultimately develops the forms of discourse most highly valued by those who have access to resources.

Tensions similar to those in the multilingual environments studied by Setati and by Barwell exist in relatively unilingual environments as well. In Chap.7, Schleppegrell explains how attention to peculiarities and specificities of mathematics discourse is worth the while of mathematics teachers. Herbel-Eisenmann has worked with teachers to explore their discourse and reports in Chap. 10 on aspects of that experience. From our experiences, we know that mathematics teachers are not at first inclined to attend to their discourse, but that attention to inequities related to their discourse practices piques their interest. Nevertheless, there are significant challenges in raising and supporting such attention.

A final critique of the access perspective is provided by Martin (2009), who states that a focus on access ignores the racialized nature of society and the work place, a perspective that could be expanded beyond race to consider other ways of framing students' identities that are associated with marginalized statuses. He claims that even when students of colour in the U.S. develop competence in high-valued practices, they are often denied the same opportunity to jobs and other resources as similarly qualified White students. Consequently, Martin argues, if educational policy ignores the ways race is constructed inside and outside of educational settings, these policies are likely to have little impact on whether or not outcomes are equitable.

3 Tensions Between the Dominant and Critical Axes

The discussion of the tension along the two dimensions of the dominant axis was previewed above in the ways the classroom-based researchers described the importance of attending to students' linguistic, cultural and political contexts and the resources that derive from those contexts. This tension exemplifies the orthogonal relationship between the dominant and critical axes. Concerns for access and achievement (the dominant axis) frame equity in terms of privileging access to high-value knowledge. Concerns for student identity and power (the critical axis) frame equity in terms of privileging access to culturally based ways of thinking and acting.

The identity and power perspective is conceived in terms of designing discourse practices around students' linguistic and cultural resources both from the perspective of the culture they bring to the classroom, which Vithal and Skovsmose (1997) refer to as students' *background*, and the perspective of looking forward to the cultures they would want to engage with in the future, which they label as the students' *foreground.* In her chapter, Gutiérrez similarly states that, in order for students to understand dominant mathematics, they should not have to divorce themselves from their current ways of being in the world. In the following subsections, we focus more explicitly on the role of these resources in providing opportunities for students to learn mathematics. We explore the tensions between culturally-based and dominant forms of discourse in mathematics classrooms, integrating the perspectives of the classroom-based researchers and policy experts who spoke to this tension. We then explore the power dimension of policy as discussed by our policy experts.

3.1 Tensions Between the Culturally-Based and Dominant Forms of Discourse

Questions raised by the classroom-based researchers surrounding the tension between culturally or linguistically based forms and dominant forms include whether these discourse forms are mutually exclusive, whether each can be developed in service of the other, whether teachers can develop the capacity to understand multiple forms of discourse and how teachers attend to issues of power and politics around classroom discourse.

Moschkovich warns that there is a potentially dangerous dichotomy in this tension:

> In terms of theory, if the study of learning and teaching for learners from non-dominant groups is relegated to being only about that group, the study of learning and teaching (writ large) will continue to assume that there is a norm (regular folk, meaning those from dominant groups) and to reflect only the experiences of learners from dominant communities. [...] In terms of practice, this perception assumes that learners from non-dominant communities *are* the problem, because they learn in fundamentally different ways than regular folk, that teaching them requires special pedagogical tricks and that we cannot learn much about how regular folk learn (or how we should teach) from our work with learners from non-dominant communities. (p. 91, 90)

As noted above, Setati and other researchers whose work is based in multilingual mathematics classrooms (e.g. Moschkovich 2002) show how learners develop an understanding of mathematical concepts while working in their first language and use English formally to communicate their results. These students engage in *code-switching*, an important practice for meaning-making, not only between languages, but also between registers within the same language to negotiate meaning. This research suggests that culturally based discursive forms *help* learners develop mathematical understanding and, ultimately, to participate in conventional disciplinary forms such as argumentation, a phenomenon that Setati labels *cognitive access*, in contrast to the *social access* emphasized by the teachers she interviewed.

The tension is particularly relevant in contexts in which learners are from marginalized backgrounds, because there is a cost to ignoring culturally based forms of discourse, a cost that Jorgensen (this volume), drawing on Bourdieu, describes as *symbolic violence*. Three chapters in this volume – by Setati, by Jorgensen and by Wagner and Lunney Borden – address the experiences of students from such marginalized backgrounds, by connecting mathematics experiences to their cultural identities. Even within such approaches, we note power implications. These power implications are particularly significant in mathematics classrooms because of the high value society ascribes to mathematics as a discipline. Wagner and Lunney Borden point out in their chapter that there are differences between teachers and other authorities identifying cultural connections on the one hand and students themselves identifying them on the other. Nevertheless, even if students identify the connections, it is important that their teachers and community leaders support the students in making these connections.

Secada also described for us the consequences of ignoring the cultural perspectives implicit in disciplinary-based forms of reasoning advocated by mathematics educators: "The kinds of argumentation that math educators promote are very culturally laden things, that make sense only for particular kids from middle-class backgrounds". He explained the implications of advocating such practices in classrooms without regard for students' cultural perspectives:

> To say that the value of an idea is based on its ability to compel someone to agree to it on the basis of its intellectual coherence and the quality of the argument, flies in the face of people and children who are socialized into saying that the value of an idea is based on who said it... Kids who are raised in traditional ways – that the parents say things and that there are no questions asked, you just obey – to place them into the settings, either makes it seem like, 'Why are you playing games with me when you [the adult] know the answer,' [...] or if you succeed in socializing them and then they transfer that way of looking at the world at home, [it might lead to tensions in the home].

Though discourse practices associated with the discipline of mathematics are based on argumentation (Forman 2003) and rely on the authority of evidence or logic to support claims, they may stand in contrast to cultural perspectives. Teachers who ignore such differences place the burden on students to understand the situated nature of such discourses and the potential risks to thinking of them as appropriate for settings outside the classroom.

4 Distribution of Power Within Educational Systems

We have not yet explored the specific structure of power in educational systems, which was a topic raised for us by some of the policy experts. In this section, we explore how power relations are evident in educational systems and connect power relations to the structure of educational systems. Doug Willms used the term "loosely coupled system" to describe the nature of policy in mathematics education. This characterization of the structure of the system describes how educational systems often do not function as formal hierarchies, despite the existence of hierarchical structures.

In his ground-breaking depiction of education structures, Weick (1976) introduced the idea of loosely coupled systems and described them as resilient to change because so many people at so many levels are active agents in the system. Most are not limited by terms of office, thus their beliefs and practices endure, surviving regime change. Weick also described how such distribution of power makes a loosely coupled system sensitive to local issues. He called this a system that perceives well.

With people in power at all levels of the system, it is possible to respond to local differences in culture, for example. Weick's analysis, and most of the comments made by the policy experts with whom we had conversations, seemed to focus on the sometimes explicit and sometimes implicit contract between mathematics teachers and jurisdictional leaders. However, from our perspective as researchers in classrooms, we notice more agents in the system. In particular, students are agents, not mere products of the system. Furthermore, the jurisdictional leaders are not the zenith in the system. We ask where their ideas and mandate come from. They too are responsive agents. The question is this: *To whom or to what are they responding?*

Regarding the 'top' end of the system, our conversation with Willms was instructive. Because he has worked extensively with high-level administrators from numerous countries, we asked him what prompts education policymakers to address issues relating to mathematics education and what prompts them to address issues in specific ways. In particular, we were interested in his views on the way policy draws on research. He answered with an account of a recent experience. When leading an international consultation of policymakers, he "had this policy group to try to get the group to formulate their policy questions, and they're actually not very good at it. [...] They don't even know what they want to know".

Willms said that the policymakers seemed unsure what change was necessary in their milieux. It was important for them to initiate positive changes, but they were not sure how to go about this. Willms said that policy often ignores clear research. Often one idea from research somehow engages an important policy leader who then pushes reform that connects with this one idea. It is unclear how researchers could promote their work to be instrumental in this way for policy.

Willms' approach has been to simplify results to catch the attention of policymakers at all levels: "I've had good success when I can just get one kind of clear idea and hammer it to death." For example, he has promoted the "shift from learning to read to reading to learn" – the idea that early literacy is necessary for achievement in all disciplines later on. He even promoted this idea in our conversation, saying that the most important way to promote equity in mathematics classrooms is to focus attention on developing reading skills in general in the first years of schooling: when students are unable to participate in discourse, which requires reading, inequities abound. Willms' use of the metaphor of the hammer in the quotation above invokes the image of policy being a blunt instrument.

4.1 Reciprocity in Classroom Positioning

On the 'bottom' end of the loosely coupled system, we know that relationships within any system are reciprocal. If educators try to position students in a certain way, students may comply or resist that positioning (Wagner and Herbel-Eisenmann 2009). Thus, students have significant power over the discourse forms that take shape in any classroom. Collectively, the response of students in numerous classrooms combines to shape teachers' views about how to position themselves. This is true whether or not policy takes students' views seriously; it is true whether or not policies set up forums for students to inform the development of curriculum (as in the case described by Paine and mentioned in Sect. 2 of this chapter).

In addition to shaping the discourse, students have the power to opt in or out of the particular discourse that presents itself to them in mathematics class. In the chapters by Jorgensen and by Wagner and Lunney Borden, they confront milieux in which mainstream mathematics education does not engage the children. Both chapters aim to redress that disparity. Children not engaging with mathematics in the classroom may be the strongest form of resistance. The success of the system is affected by students engaging or disengaging with the particular discourse of mathematics they face in classrooms.

With our interest in equitable discourse, we see the benefits of a loosely coupled education system. We suggest that these benefits would best be realized if the distribution of agency were recognized at all levels of the system. As in the examples in China given by Paine (see Sect. 2), not only were teachers and students agents in the system, but the system was set up to recognize their agency as people who knew what they needed to improve teaching and learning and to adapt to their voices. Indeed, the recommendations made by Moschkovich in Chap. 6 would require the attention of agents in all levels of the system.

Not only do students interact with teachers and thus shape the discourse within the loosely coupled mathematics education system, but students also interact with each other and thus open or close space for each other within the system. Esmonde's chapter describes some of this power dynamic. So students, who are the majority agents in the system, exercise power over teacher–student discourse, over the success of the system and over each other.

4.2 Educational Systems as a Web of Relations

The loosely coupled system of mathematics schooling is coupled in yet further ways beyond the contract between policy and mathematics teachers. Though Weick focused on the hierarchy comprising administrators, teachers and students, he also noted that "under conditions of loose coupling one should see considerable effort devoted to constructing social reality, a great amount of face work and linguistic work, numerous myths" (p. 13). This advice aligns with Wagner and Herbel-Eisenmann's (2009) call for the remythologizing of mathematics education, as they drew attention to the way people are positioned within mathematics classrooms, as an alternative to envisioning students as merely respondent to a hierarchical discipline coming from outside their classroom walls.

Weick's recognition of the significance of social reality points to the connections among mathematics policy, mathematics classroom events and other cultural phenomena, which connects in various ways to mathematics and school. Weick described how loosely coupled systems are resilient because change requires changes in beliefs and values among all the agents at play in the system (which we are seeing as far-reaching). For example, change might require contending with issues of privilege and oppression throughout society. These kinds of far-reaching connections are likely to require attention, in order, for example, for the disparities in mathematics achievement along socioeconomic lines in the U.S.

The case of Bhutan's reforms in mathematics education shed light on this kind of dynamic at work in systemic reform. Starting from 2006, Bhutan's government implemented a gradual change in the curriculum. The most significant recommendation involved classroom discourse structure. Karma Yeshey described the upshot of the change in this way:

> Perhaps one of the most important aspects of the new curriculum is the requirement to explain and communicate ideas and understandings; to provide justification and reasoning for the solutions consistently. The teacher will need to be all-supportive in this. Over time then, the students will, hopefully ask the teachers too to give reason and justification for his/her own answers. We earnestly need to promote this in our teachers and students in Bhutan, for in Bhutan we have somehow the culture of not questioning teachers and elders. This will improve the powers of our rational and critical minds!

Yeshey recognized the need for both teachers and students to recognize the need for altered discourse forms in mathematics class. Our claim that other social phenomena connect with the mathematics education system is quite clearly pertinent in the case of Bhutan, for the country is simultaneously undergoing massive reform in government, moving from a monarchy to a democracy. It is already described as a democracy, but leadership recognizes that this is a slow process.

The reform of mathematics is intentionally connected to the development of democracy. Yeshey said, "I think Democracy will thrive only if we have a questioning and reasoning people and an explaining and reasoning government." Because the system of mathematics education is loosely coupled itself and coupled, albeit loosely, to other social forces in the surrounding culture, democracy depends on mathematics education as much as discourse in mathematics education depends on students' conceptions of appropriate ways of interacting with teachers and elders.

The concurrent reform of mathematics education and political structure draws attention to the need for policy change in mathematics education to be responsive to cultures at work in the communities in which schools are situated. Relatively global cultural phenomena (as in the national reforms in Bhutan) and relatively local phenomena are already connected to what happens in the mathematics classroom. Thus, it is important for policymakers at all levels to take this connection seriously.

At the relatively global level, policy might prescribe changes that align with parallel cultural changes (as in Bhutan). The relatively local-level connection between mathematics classroom experience and students' cultural milieux is central to Gutiérrez' identification of identity as an aspect of equity in mathematics education. As Weick noted, the loosely coupled nature of education makes it possible for teachers to be responsive to local particularities. We note that large-scale policy can recognize the need for teachers to be responsive to local culture, and can structure systems that support connections between mathematics classrooms and community experiences. Wagner and Lunney Borden's chapter describes one instance of a structure that promotes such local interaction.

5 Dilemmas of Moving Forward

The policy experts and researchers described dilemmas related to policy that aims to provide access to dominant forms of mathematics. These communities described the difficulty of being sensitive to local contexts while at the same time providing opportunities for students to understand and master dominant mathematical forms of language and reasoning. On the one hand, the policy experts described the limited reach of policy that strictly mandates particular practices, without providing the autonomy and resources to develop the capacity to enact challenging forms of curriculum and instruction locally. On the other, the researchers described the perils of ignoring the kinds of local linguistic forms (e.g. the use of multiple languages and informal genres) that provide cognitive access to mathematics, while at the same time warning of the risks of highlighting these linguistic forms in terms of maintaining the marginalized status of those most likely to use multiple languages and informal genres.

There is, however, some potential convergence in the two communities, particularly around the notion of building capacity to enact challenging forms of curriculum and instruction across a broad array of contexts. One possible site of convergence stems from Paine's example of the role of policy in China to provide support to schools that are struggling to help their students learn mathematics. She described for us the ways that local schools are provided both autonomy and support to meet the demands of teaching challenging concepts to a broad range of students. King would add that such locally oriented policies should align the perspectives of educators with those of the local community. The classroom-based researchers might contribute to this idea by inquiring into the kinds of practices in which teachers should develop capacity, with the idea of highlighting how access to particular practices leads to achievement through first providing cognitive or linguistic access. Current policies, they argue, are misguided in their fundamental understanding of how language use is associated with learning and should promote the effective use of multilingual resources and informal genres.

The prescriptive policy focused on measuring student learning described above addresses the dominant axis, because it measures success of the policy in terms of large-scale measures such as access and achievement. By contrast, policy that structures support for teachers as agents can address explicitly the need for them to be responsive to the identities of students and to foster the power of students and the community in classroom dynamics. Identity and power comprise the critical axis in Gutiérrez' account. However, because education systems are by nature loosely coupled, the reality is that teachers are free to be locally responsive.

Nevertheless, they may not be inclined or equipped to be responsive. Furthermore, even when policies go beyond prescription to capacity building, the question remains 'Building capacity to do what exactly?' If it is about building capacity to increase achievement and access, such policy in effect does not speak to the critical axis in any meaningful way. This, perhaps, reflects the politically risky dimension of this axis and the difficulties of taking context and culture into consideration when crafting and implementing policy at a large scale. One way for research on equity and discourse to make the critical axis more relevant in mathematics classrooms is to illustrate how issues of identity and power impact mathematics achievement, for example, by taking Moschkovich's recommendations for research seriously. However, this imperative must be mitigated by modifications of the system that privileges certain cultures.

The tension between access and identity in educational policy and practice is not easily managed, much less resolved. Three issues emerge. First, is it possible to mandate the use of non-dominant or non-disciplinary-based discourse forms? Martin (2009) is sceptical on this point, observing that "top-down, externally generated solutions that are not responsive to the needs and conditions of the context in question are unlikely to have a meaningful and lasting effect" (p. 304). We argue, however, that change is possible if teachers understand connections between suggested new discourse forms and their own views of social needs or social justice.

Second, even if teachers began to encourage the use of such linguistic moves as code-switching, would they be able to understand and navigate the political implications of these moves? Given the often-reactionary backlash to progressive reforms (Tyack and Cuban 1995), teachers may lack the political will or cultural capital to withstand criticism of such practices, as noted by Setati. Third, from a strictly pedagogic perspective, how can teachers help students to understand the relationships between culturally-based forms of discourse and those 'high-value' forms that offer access to resources in the dominant society? Herbel-Eisenmann's work with teachers described in Chap. 10 shows that, even as teachers engage seriously with issues related to taking up new classroom discourse practices, this might only scratch the surface of coming to understand the discourse practices of all of the other communities in which students participate in order to consider how to negotiate the hybridity of these practices.

Policy as blunt force can at best mandate a change in the official discourse; the translation to practice is questionable. *How can policy be sensitive to local conditions in ways that spur change in practices? How can policy impact the way race and language use are construed?*

Two themes emerged from our analysis relative to possibilities. The first is that a teacher-centred focus has possibilities, especially if teachers engage in inquiry about their students' cognitive, linguistic, cultural and political resources. In this vein, Setati argues in Chap. 8 that teachers need to take a holistic view of learners:

> Multilingual learners have a unique and specific language configuration and therefore they should not be considered as the sum of two or more complete or incomplete monolinguals. The possible solution to the problem explored in this chapter is informed by this holistic view of multilingual learners. (p. 133)

Barwell similarly argues that policy needs to move away from deficit notions of students that place the problems with participation on students' lack of understanding of terminology or register. Instead, he suggests, policy should at least recognize the complexities involved in asking students to engage in mathematical discourse in classrooms, especially when they are simultaneously learning the language of instruction.

We have also become more aware of macro-cultural differences in dealing with and implementing change. Ironically, China is often portrayed as a relatively authoritarian culture and the U.S.A. as relatively democratic; yet, the nature of policy in education seems to be the opposite of what we would expect. Clearly, the relationship between culture and policy is complex.

The case of Bhutan's mathematics curriculum change provides further insight into the role of macro-culture. As quoted above, Yeshey characterized the Bhutanese view of authority in terms of policy in this way: "In Bhutan, we have somehow the culture of not questioning teachers and elders". This quotation demonstrates his awareness that policy change initiatives are, in fact, attempts to change culture (and power dynamics), even while guided by culture. Later in the interview, he referred to the culture as evolving. Seeing culture as dynamic and power-laden is important in the consideration of change.

We offer these suggestions in attempting to address Morgan's question posed at the start of this chapter: "*What* forms of pedagogy might prepare *which* students to participate in [critical and transformative action]?" The discussion in this chapter suggests that policy cannot prescribe practices, especially in technical terms, but should recognize the complexities involved in engaging students in mathematical discourse and should support teachers in inquiring into their students' practices and ways of reasoning.

Afterword
Six Post-Its in Search of an Author

THE FATHER: "But don't you see that the whole trouble lies here. In words, words."

(Pirandello 1922/1931, p. 16)

I gave the first 15-min talk as part of the opening panel at the May 2008 conference that comprised the genesis of this volume, as well as provided its focus. Here, nearly 3 years later, I am writing its closing words. My talk in 2008 was not as well thought through as it should have been, so most of it will remain unwritten. But I did start with some words about democracy (see Sinclair and Pimm 2010), a word notable by its absence from this book (mentioned only a handful of times, all but one in the final chapter).

This is not a systematic attempt to look back across this book as a whole (reviewers can try their hand at that), nor is it a review of the area it dwells in. The book-end Chapters 1 and 13 have undertaken this latter task, whose authors were the primary organisers and conceptualisers of both the Rochester conference and this volume. My closing series of notes (metaphorically small enough that each one could fit on a Post-it) structurally bookends Jill Adler's *Foreword* (she nobly at the prow, me so much baggage at the stern). But unlike her piece, which frames things broadly with one or two specific, subsequent chapter or thematic mentions, mine will start with specifics gleaned from some of the foregoing chapters, striking details that caught my attention or interesting ideas that caused me to stop and think, and then orient itself outward, back into the world.

Post-its act as reminders, breadcrumbs left in the world like knots tied tightly in string or in handkerchiefs, in order to trigger a memory, one that, it is often feared, might otherwise go missing or be forgotten. I offer my half-dozen[1] here in that same

[1]Revisiting Luigi Pirandello's striking play about language, reality and identity, *Six Characters in Search of an Author*, provided me with a starting point for this piece. But the number six also crops up in two other notable Italianate works: Italo Calvino's (1988/1992) *Six Memos for the Next Millennium* and Umberto Eco's (1994) *Six Walks in the Fictional Woods*. In Pirandello, two of the six disembodied, eponymous characters never speak while Calvino sadly died before completing his sixth memo (on 'consistency'). Only Eco provides us with the full half-dozen promised in his title.

B. Herbel-Eisenmann et al. (eds.), *Equity in Discourse for Mathematics Education: Theories, Practices, and Policies*, Mathematics Education Library 55,
DOI 10.1007/978-94-007-2813-4, © Springer Science+Business Media B.V. 2012

spirit. They act for me as things I want to remember to think more about. But one nice thing about these yellow stickies is that they can adhere to anything – even to other Post-its. So this is a warning that there is to be no nice numbered list: you are, of course, welcome to read them in any order.

The contemporary European historian Tony Judt died in New York from ALS in August 2010. A little earlier that same year, when asked by a friend and former student about an epitaph, Judt proposed, "I did words" (Goldberg 2010). I, too, have done words for the past 35 years, both producing them and studying them in their mathematical and educational contexts. 'Doing words', for me, is a way of acknowledging their power and the singular sway they have held for me throughout my life. "At the bottom of each word/I'm a spectator at my birth", wrote poet Alain Bosquet. As I write this afterword, Kai, my 16-month-old grandson, plays at my feet,[2] birthing his words as best he can, words to which I shall return.

Post-It Time

In Chap. 8, Setati identified two central principles for her proposal, one to do with deliberate, explicit use of learners' home languages and the other to do with the use of challenging real-world mathematical tasks. A possible note might be to consider the strength of interaction between these two principles, as opposed their simple concatenation. However, what particularly caught my attention was the sense of history repeating itself with regard to arguments concerning the mathematics register (bearing in mind Santayana's often misquoted pseudo-admonition, that "those who cannot remember the past are condemned to repeat it").

The English language has not always been so seemingly monolithic and apparently hegemonic (through the extent to which it seems temporarily so global today may be somewhat overstated). Less than 500 years ago, the English mathematics register did not exist. The first book on arithmetic written in English was published in 1537. Academic discussion in much of Europe took place in Latin and, to a lesser extent, in Greek. English 'grammar' schools were 'Latin grammar' schools, those that provided access to this apparent language of the intellect.

In the sixteenth century, there was a major dispute in England about the fitness of the vernacular (English) to support academic discussions. Fauvel (1987) observes:

> There was much debate, in which many humanist scholars were involved, over the appropriateness and possibility of writing down matters of any subtlety or technicality into English. Discussions about translating classical texts into the vernacular – and, indeed, about writing texts in the vernacular in the first place – are found throughout the century. (p. 10)

[2]Like the later Freudenthal, perhaps, I am in danger of entering my anecdotage.

He nicely documents some of the arguments on both sides, including a poem by John Skelton from around 1504 arguing English's deficiencies as a satisfactory tool of epistemological expression:

Our natural tong[ue] is rude
And hard to be ennewed
With polysshed terms lusty;
Our language is so rusty [...]
I wot not where to find
Terms to serve my mynd.

and some lines from a play by John Rastell, "popularising astronomy and cosmography which urged the English language be used for science" (p. 11):

Then if cunning latin books were translate
In to english well correct and approbate
All subtle science in english might be learned
As well as other people in their own tongues did.

Fauvel then proceeds to identify Robert Record as a major mathematics textbook author of the mid-sixteenth century attempting single-handedly along the way to create a mathematics register for English *based on Anglo-Saxon rather than Latin or Greek roots*. Unfortunately, this home-grown register did not take in the main: the so-called English mathematics register is mostly grafted from a mix of Latin and Greek word-stock.

You can likely tell I am already on at least my second Post-it note by this stage, it hanging from the first. My intent is to signal that some useful and delicate diachronic work in mathematics education could be to document how the mathematics registers of various European (and other) languages came to be, in order to learn more from history about such register creation. In passing (in small print running up the side perhaps), Setati cites in passing Benton's (1978) work identifying the project of formal schooling for the Maori community as being about providing access to English, yet we know from Barton et al. (1998), a mere 20 years later, that a register for school mathematics to grade 12 had by then been created: not only created, but in use in certain Maori schools as a consequence of a rapid, systematic and conscious development by a small but influential group of people. Even 15 years further on as we almost are now, a follow-up report on the present state of affairs would be highly informative (and, of course, such a document may well exist and I am simply unaware of it – see also Barton 2008)[3].

In South Africa, however, with nine official African languages that *each* might develop further a mathematics register suited to school-level mathematics, this difference in quantity (as well as the possible significant unease at privileging one, perhaps, as being the next alongside English or Afrikaans, for mathematics at least) may not make this a viable path: it may become a difference in quality. But one thing I take from the history of English and contemporary developments in Maori is how *relatively* quickly a position, seemingly a fixity or a taken-for-granted, can change drastically (see also Roberts 1998).

[3]With an eerie timeliness, I have just obtained a copy of Meaney, Trinick and Fairhall (2012), which does exactly that.

My Post-it chain is, in some ways, of quite a different order, but also connects indirectly to the mathematics register, a touchstone to which I keep returning. It arose initially from Chap. 2 with Gutiérrez' helpful and significant expansion of the notion of equity into four … four whats? She refers initially to them as 'dimensions', but then proceeds to pair them into what she terms the *dominant* and *critical* axes. As I had previously envisaged these dimensions as individually lying along axes, this tripped my Cartesian attention. Moschkovich, in her chapter, spoke of interactions among individual elements from contrary pairs, while in the final chapter Choppin, Wagner and Herbel-Eisenmann write of the relation between these two axes as being 'orthogonal' (p. 215). And, albeit in quite a different context, Barwell, in Chap. 9, offers us Cummins' diagram involving Cartesian axes (Fig. 9.1) and then writes of the considerable degree of interdependence between the two dimensions of cognitive demand and degree of context.

I may well be suffering from an excess of Linear Algebra (ironically, the only other place where the word 'orthogonal' arises is in Laghribi's mathematical text briefly cited in Chap. 8), but the mathematical notions of dependence and independence collided for me with that of interdependence. It is a familiar state of affairs where semantic contamination (or, here, cross-contamination) occurs across registers: but it is still often a surprise to me when a new instance crops up. So this Post-it sequence revolves around a reminder for me to look out for further instances of this semantic collision, as well as to look for more ways (likely from *outside* mathematics) to talk about interdependence of component factors.

Much in this book has addressed various forms of community and the specificity or otherwise of various forms of discourse and discourse practices (e.g. around questions). But the school can also be considered a community,[4] one in which many children spend the majority of their waking hours. There may well be a greater or lesser discord with certain discourse practices upon entering the school, but some of them are there for specific purposes. Just as development of a mathematics register perturbs the nature of the grammar (a significant complaint made by Maori speakers[5]), so do the nature and demands of schooling regularly violate the Gricean maxims of conversation.

School is an artificial setting, one intended to offer heightened experiences, experiences that are not available outside. This is especially true of mathematics classrooms (Pimm 1994). What is less well worked out is how forms relate to functions, to the intents of teaching mathematics and the forms, traditional and not yet so, in which that enterprise has, does and will continue to come wrapped.

[4]Adler (1998b) has written tellingly about the mismatch between certain attempts to overlay Lave and Wenger's community of practice notions onto the classroom (the students are not apprenticed to the teacher, there is no common practice into which they are being apprenticed, the teacher is not usually also a practitioner, …).

[5]This fact should not be taken lightly. To perturb something as significant and stable as a language grammar is an indicator of significant semantic pressure. Precisely why would be a worthwhile on-going exploration.

Adler's (2001) three central dilemmas are linked both to each other and to a fundamental aspect of language, its transparency. How does teaching fundamentally make use of that fact in order to achieve its ends?

Transparency is there with technology, as Setati points out. It is there with individual symbols: if they cannot become invisible, then they cannot function symbolically. But, at others times, they must be visible, in order to be formed and attended to. Mathematics makes use of connections between visible symbols and aspects of the things symbolized (it is what makes symbol manipulation possible and fluency requires this: see Tahta 1991 and Pimm 1995). Mathematics itself might, in this sense, be seen as a (the?) transparent discipline.

More Post-its: they are coming thick and fast now. Barwell's attention in Chap. 9 to conversation analysis, and Sacks' observation that turn-taking as a fundamental phenomenon of conversation, triggered the thought that in school 'conversations' it is often turn-*giving* (by the teacher) as much as turn-*taking* (by the student). I wondered what effects this might have on the structure of the event and on the fact that the word 'demand' in his term *discursive demand* might have particular resonance in such settings.

On gaining access to a mathematical identity: year after year, students all over the world, both in *soi disant* advanced economies as well as in developing ones, are refusing to take on mathematical identities. At what point do I, as a student, have that right, knowingly to walk away from one, even if the economic and social effects of so doing can be significant? At the other extreme, can I maintain or develop my mathematical identity (and at what cost, both material and psychic) in a school context that may be inimical to this?

A final note for now has to do with mathematics itself? 'The motley of mathematics' was a courageous phrase of Wittgenstein's penned precisely at the time the high structuralism of Bourbaki was emerging in the inter-war period. This was what, I felt, was indirectly being drawn on in this book at times, an echo of Moschkovich's felt and strong opening observation about her work being framed for her by its interaction or otherwise with marginalized communities. How is mathematics referred to here in this book and does it tend to slip away or otherwise absent itself, when discourse is in focus? Is it treated as a monolithic entity rather than a motley?

Halliday, Language and Making the World

> Languages have different patterns of meaning – different 'semantic structures' in the terminology of linguistics. These are significant for the ways their speakers interact with one another; not in the sense that they determine the ways in which the members of the community *perceive* the world around them, but in the sense that they determine what the members of the community *attend to*. (Halliday 1978, p. 198)

I keep coming back to Michael Halliday and his rich, subtle view of the significance, force and essential mutuality of language and human language use, as well as his lifetime encounter with examining closely what gets made when a child makes her or his own language. In this book, Schelppegrell's thorough laying out in Chap. 7 of what even some elements of SFL can offer mathematics education is the most evident trace of Halliday's influence in this book (and hers is a chapter that has disappeared under a mass of yellow squares). But there are other places too where Halliday kept surfacing, to my mind sometimes equipped with a gentle, cautionary finger.

But my chosen genre for this piece is not that of a book reviewer, but more one of an AERA symposium discussant. So, instead, having run out of Post-its, though not out of places to put them, I end by returning to Halliday's (1975b) work *Learning How to Mean*[6] and, indirectly, to Kai, my constant and far-from-silent companion.

Halliday's book includes many close observations of his young son coming to language and, in particular, how the process involves "generalization followed by abstraction". (This was something to which Caleb Gattegno was highly attuned in his view of language learning as involving a high degree of mathematisation: indeed, it might represent the human child's first significant encounter with the central processes of mathematics.) But also Halliday distinguishes the *mathetic* from the *pragmatic* functions in infants learning language, glossing the difference as "the use of the symbolic system not as a means of acting upon reality but as a means of learning about reality" (p. 106). For far too many people, mathematics and, in particular, its forms of language serve neither function.

David Pimm

[6]Arguably, his choice of title single-handedly altered the grammatical category of this specific verb from 'transitive only' to 'transitive or intransitive'.

Contributing Authors

Jill Adler (jill.adler@wits.ac.za) holds the FRF Mathematics Education Chair at the University of the Witwatersrand, South Africa, and the Chair of Mathematics Education at King's College London, U.K. She is also a Fellow of ASSA, the Academy of Science of South Africa. She leads the QUANTUM research project, the focus of which is *Mathematics in and for Teaching*, and the research and development projects within the FRF Chair. Professor Adler is past Vice-President of the International Commission for Mathematical Instruction (ICMI), and recipient of both the Vice Chancellor's Research Award, and the Vice Chancellor's Team Award for Academic Citizenship at Wits University. She is the author and/or editor of three books, on teaching and learning mathematics in multilingual classrooms, professional development and mathematics education research in South Africa.

Celia Rousseau Anderson (croussea@memphis.edu) is an associate professor of mathematics education at the University of Memphis, U.S.A. Her primary research areas include equity and opportunity to learn in mathematics education and teacher professional development. Dr. Anderson is currently the PI of the Memphis Mathematics and Science Teacher Induction Fellowship (a Robert Noyce Fellowship grant) and Co-Director of the UTeach Memphis program (a UTeach replication). She has publications in *For the Learning of Mathematics*, *School Science and Mathematics*, the *Journal of Urban Mathematics Education*, an NCTM yearbook and the *Handbook of International Research in Mathematics Education*.

Richard Barwell (rbarwell@uottawa.ca) is an associate professor in the Faculty of Education, University of Ottawa, Canada. His research lies at the intersection of mathematics education and applied linguistics, and he has published in peer-reviewed journals in both domains. His research interests include bilingual, multilingual and second language learners of mathematics, the nature of mathematics classroom discourse and the use of discourse analysis in researching mathematics learning. He has guest edited special issues on these topics in journals such as *Educational Studies in Mathematics* and *Linguistics and Education*. He recently edited the book *Multilingualism in Mathematics Classrooms: Global Perspectives* (published by Multilingual Matters) and is the current editor of *For the Learning of Mathematics*.

B. Herbel-Eisenmann et al. (eds.), *Equity in Discourse for Mathematics Education: Theories, Practices, and Policies*, Mathematics Education Library 55,
DOI 10.1007/978-94-007-2813-4, © Springer Science+Business Media B.V. 2012

Jeffrey Choppin (jchoppin@warner.rochester.edu) is an associate professor of mathematics education at the University of Rochester, U.S.A. His current research focuses on teachers' conceptions of how innovative curriculum materials enable potential learning trajectories. He is interested in how learning opportunities are distributed within and across classrooms, particularly with respect to mathematical discourse practices. His current projects include the NSF-funded ACCLIME (Adapting Curriculum for Learning in Mathematics Education) project. Choppin has published articles in *Curriculum Inquiry, Mathematical Thinking and Learning, Journal of Mathematics Teacher Education, Mathematics Education Research Journal, Mathematics Teacher* and the NCTM yearbooks. He taught mathematics for 12 years in the Washington, DC public schools and won the Presidential Award for Excellence in Mathematics and Science Teaching in 1995.

Indigo Esmonde (indigo.esmonde@utoronto.ca) is an assistant professor of mathematics education at the University of Toronto, Canada. Dr. Esmonde's research focuses on issues of equity in diverse mathematics classrooms, with a specific focus on connections between racialized and gendered identities and mathematical ones. A current research project, funded by the Knowles Science Teaching Foundation, is examining teacher learning through action research on equitable mathematics education. Dr. Esmonde's research has been published in *Review of Educational Research, Journal of the Learning Sciences,* and the *Journal of Urban Mathematics Education*, and she recently co-edited a special issue of the *Canadian Journal of Science, Mathematics and Technology Education* focused on equity and discourse.

Rochelle Gutiérrez (rg1@illinois.edu) is Professor of Curriculum and Instruction and Latina/Latino Studies at the University of Illinois, Urbana-Champaign, U.S.A. Her research focuses on equity in mathematics education, race/class/language issues in teaching and learning mathematics, effective teacher communities, and the achievement gap. Her current research projects include: teacher community and secondary mathematics teaching in México (for which she received a Fulbright); developing pre-service teachers' knowledge and disposition to teach mathematics to marginalized students, and using 'Nepantla' as a way to theorize knowledge for teaching. She currently serves as guest editor for a special issue on identity/power for the *Journal for Research in Mathematics Education*. Before and throughout graduate school, she taught middle and high school mathematics to adolescents in East San José, California.

Donna M. Harris (dharris@warner.rochester.ca) is an assistant professor of educational leadership at the University of Rochester, U.S.A. Her research focuses on school reform, educational policy, and the social organization of public schools and classrooms. Specifically, Dr. Harris focuses on how educational institutions, policies, and practices affect learning opportunities and experiences – especially for students of color. Her expertise includes ability grouping and tracking, whole-school reform, standards-based education, and high-stakes testing. Harris' current research collaboration examines the school experiences of Latina/o students in Rochester,

New York. Her most recent publication in *Educational Policy* examines how curriculum differentiation occurs in schools involved with comprehensive school reform. For 2011–2013, she is chair of the Tracking/Detracking Special Interest Group of the American Educational Research Association.

Beth Herbel-Eisenmann (bhe@msu.edu), a former junior high mathematics teacher, is an associate professor of mathematics education at Michigan State University, U.S.A. Her research focuses on examining written, enacted and hidden curricula by drawing on ideas and methods from sociolinguistics and discourse literatures. She spent 5 years in collaborative research with eight secondary mathematics teachers who used action research to align more closely their discourse practices with their professed beliefs. This work was published in *Promoting Purposeful Discourse: Teacher Research in Mathematics Classrooms* (NCTM, 2009). She serves on the Editorial Board of the *Journal for Research in Mathematics Education*, the Board of Directors for the Association of Mathematics Teacher Educators, and the International Advisory Board and Organizing Committee of the *Mathematics Education and Society* conference.

Robyn Jorgensen (r.jorgensen@griffith.edu.au) is Professor of Education at Griffith University, Australia. She has worked in the area of equity and mathematics education for 2 decades. She is concerned with students from low SES backgrounds, remote and rural students as well as Indigenous students. She draws on the work of Bourdieu to theorise the ways in which practices in school mathematics work to exclude some groups of students. She has published widely and has been the Chief Investigator on 9 Australian Research Council projects. She is currently editor of the *Mathematics Education Research Journal* and on the Editorial Board of the *International Journal for Science and Mathematics Education*. She is on the Advisory Board for the Australian Association of Mathematics Teachers 'Make it Count' Project, which aims to enhance the mathematics learning for urban Indigenous students.

Lisa Lunney Borden (lborden@stfx.ca) is an Assistant Professor at St. Francis Xavier University in Canada. She teaches mostly mathematics education courses at both the undergraduate and graduate level. She began her career teaching grades 7–12 mathematics at We'koqma'q First Nation Secondary School, a Mi'kmaw community-run school. She credits her students and the Mi'kmaw community for inspiring her to think differently about mathematics education. Lisa is most interested in examining strategies to transform mathematics education for Aboriginal students with a focus on equity, diversity and the inclusion of multiple world-views. She has worked with the Nova Scotia Department of Education to review and revise provincial mathematics curricula and she continues to provide professional support for teachers throughout the province. Lisa also helps to coordinate an on-going project called "Show Me Your Math" that invites Aboriginal children in Atlantic Canada to explore the mathematics all around them. She has published in *For the Learning of Mathematics* and welcomes communication relating to her research interests.

Candia Morgan (c.morgan@ioe.ac.uk) has worked as a secondary school mathematics teacher and as a teacher educator and is currently Reader in Education at the Institute of Education, University of London, United Kingdom, and Visiting Professor at the University of Stockholm, Sweden. Her research develops and applies linguistic and discourse-analytic methods for investigating and critiquing social practices in mathematics education. She is interested in how official discourses of policy, curriculum and assessment contribute to the construction of mathematics, mathematical activity and the identities of teachers and students in mathematics classrooms. She is the author of *Writing Mathematically: The Discourse of Investigation* (Falmer 1998) and has published widely elsewhere, contributing to research journals and edited books. As well as serving as Associate Editor for *Educational Studies in Mathematics*, she is a member of the editorial board of the *Journal for Research in Mathematics Education*. Until recently, she served as a member of the International Advisory Board of the *Mathematics Education and Society* Conference and as Coordinator of the *Language and Mathematics* Working Group of the Congress of the European Society for Research in Mathematics Education.

Judit N. Moschkovich (jmoschko@ucsc.edu) is Professor of Mathematics Education in the Education Department at the University of California at Santa Cruz. In her research, she uses a sociocultural perspective to examine student understanding of linear functions, mathematical discourse, mathematical discussions among bilingual Latino/a students, and the relationship between language(s) and learning mathematics. Dr. Moschkovich was co-editor, with M. Brenner, of the NCTM monograph *Everyday and Academic Mathematics: Implications for the Classroom* (2002) and editor of *Language and Mathematics Education* (Information Age Publishing, 2010). She has published in *The Journal for the Learning Sciences, Educational Studies in Mathematics,* and *For the Learning of Mathematics*. She was the Principal Investigator of an NSF project *Mathematical discourse in bilingual settings: Learning mathematics in two languages* and one of the PIs for the NSF-funded Center for the Mathematics Education of Latinos/as (CEMELA). She has served on the Editorial Panel for the *Journal for Research in Mathematics Education* and as the Chair for the AERA (American Educational Research Association) *SIG* (Special Interest Group) *Research in Mathematics Education*. She currently serves on the Editorial Board for *The Journal for the Learning Sciences* and on the International Program Committee of the International Council for Mathematics Instruction (ICMI) Study #21: *Mathematics Education and Language Diversity*.

David Pimm (david.pimm@ualberta.ca) became professor emeritus at the University of Alberta, Canada upon his retirement in June 2010. For the previous decade, he had been a professor of mathematics education there, in the Department of Secondary Education. For most of his previous career, he worked in the Centre for Mathematics Education in the Department of Mathematics at the Open University in the U.K. For the past 35 years, he has been caught up in exploring and examining complex interrelationships between language and mathematics, primarily but not exclusively in classroom settings. Pimm has authored two books and edited another half-dozen, as well as published widely in other arenas. He was also editor of *For the Learning of Mathematics* for 6 years.

Mary Schleppegrell (mjschlep@umich.edu) is a professor of education at the University of Michigan, Ann Arbor, U.S.A. Her research draws on systemic functional linguistics to explore the relationship between language and learning. She is the author of *The Language of Schooling* (Erlbaum, 2004); and with Cecilia Colombi, *Developing Advanced Literacy in First and Second Languages: Meaning with Power* (Erlbaum, 2002), as well as *Reading in Secondary Content Areas: A Language-Based Pedagogy,* with Zhihui Fang (University of Michigan Press, 2008). Her work on the language challenges of mathematics has been published in *Reading and Writing Quarterly* (2007), NCTM's *Mathematics for Every Student: Responding to Diversity, Grades 6–8* (2008), and the recently published *Language and Mathematics Education: Multiple Perspectives and New Directions for Research,* edited by Judit Moschkovich (Information Age Publishing, 2010).

Mamokgethi Setati (setatrm@unisa.ac.za) is full professor and Executive Dean of the College of Science, Engineering and Technology at UNISA, South Africa, and honorary professor of mathematics education at the University of the Witwatersrand. Her research focuses on mathematics in multilingual classrooms and she is currently exploring innovative pedagogies for multilingual mathematics classrooms. She is co-chair of ICMI Study 21 on Mathematics and Language Diversity. She is former President of the Association for Mathematics Education of South Africa (AMESA), former Chairperson of the Board of the South African Mathematics Foundation (SAMF) and current Secretary of the African Mathematics Union Commission for Mathematics Education in Africa. She is a member of the Academy of Science of South Africa and an honorary member of the Golden Key International Honour Society.

David Wagner (dwagner@unb.edu) is associate dean and a mathematics education associate professor in the Faculty of Education of the University of New Brunswick, Canada. He is most interested in human interaction in mathematics and mathematics learning and the relationship between such interaction and social justice. This inspires his research which has focused on identifying positioning structures in mathematics classrooms by analyzing language practices, on ethnomathematical conversations in Mi'kmaw communities and on working with teachers to interrogate authority structures in their classrooms. He currently serves as managing editor on the board of directors of the journal *For the Learning of Mathematics* and as a member of the Nonkilling Science and Technology Research Committee. He has taught grades 7–12 mathematics in Canada and Swaziland. He has also taught mathematics education courses in Bhutan, Thailand and Trinidad.

References

Adams, B. (2009). *Owning professional development: The power of teacher research*. Unpublished Ph.D. dissertation. Iowa State University, Ames.

Adler, J. (1998a). A language of teaching dilemmas: Unlocking the complex multilingual secondary mathematics classroom. *For the Learning of Mathematics, 18*(1), 24–33.

Adler, J. (1998b). Lights and limits: Recontextualising Lave and Wenger to theorise knowledge of teaching and of learning school mathematics. In A. Watson (Ed.), *Situated cognition and the learning of mathematics* (pp. 161–177). Oxford: University of Oxford Department of Educational Studies, Centre for Mathematics Education Research.

Adler, J. (1999). The dilemma of transparency: Seeing and seeing through talk in the mathematics classroom. *Journal for Research in Mathematics Education, 30*(1), 47–64.

Adler, J. (2001). *Teaching mathematics in multilingual classrooms*. Dordrecht: Kluwer Academic Publishers.

AERA. (2006). Do the math: Cognitive demand makes a difference. *Research Points, 4*(2), 1–4.

Aikenhead, G. (2002). Cross-cultural science teaching: *Rekindling Traditions* for Aboriginal students. *Canadian Journal of Science, Mathematics and Technology Education, 2*(3), 287–304.

Antil, L., Jenkins, J., Wayne, S., & Vadasy, P. (1998). Cooperative learning: Prevalence, conceptualizations, and the relation between research and practice. *American Educational Research Journal, 35*(3), 419–454.

Atweh, B., Bleicher, R., & Cooper, T. (1998). The construction of the social context of mathematics classrooms: A sociolinguistic analysis. *Journal for Research in Mathematics Education, 29*(1), 63–82.

Atweh, B., Forgasz, H., & Nebres, B. (2001). *Sociocultural research on mathematics education: An international perspective*. Mahwah: Lawrence Erlbaum Associates.

Au, K. (1980). Participation structures in a reading lesson with Hawaiian children: Analysis of a culturally appropriate instructional event. *Anthropology and Education Quarterly, 11*(2), 91–115.

Au, K., & Jordan, C. (1981). Teaching reading to Hawaiian children: Finding a culturally appropriate solution. In H. Trueba, G. Guthrie, & K. Au (Eds.), *Culture in the bilingual classroom: Studies in classroom ethnography* (pp. 139–152). Rowley: Newbury House Publishers.

Austin, J., & Howson, A. (1979). Language and mathematical education. *Educational Studies in Mathematics, 10*(2), 161–197.

Azmitia, M. (1988). Peer interaction and problem solving: When are two heads better than one? *Child Development, 59*(1), 87–96.

Ball, D. (1988). Unlearning to teach mathematics. *For the Learning of Mathematics, 8*(1), 40–48.

Ball, D. (1993). With an eye on the mathematics horizon: Dilemmas of teaching elementary school mathematics. *The Elementary School Journal, 93*(4), 373–397.

B. Herbel-Eisenmann et al. (eds.), *Equity in Discourse for Mathematics Education: Theories, Practices, and Policies*, Mathematics Education Library 55,
DOI 10.1007/978-94-007-2813-4,

Ball, D., Thames, M., & Phelps, G. (2008). Content knowledge for teaching: What makes it special? *Journal of Teacher Education, 59*(5), 389–407.

Barron, B. (2000). Achieving coordination in collaborative problem-solving groups. *The Journal of the Learning Sciences, 9*(4), 403–436.

Barron, B. (2003). When smart groups fail. *The Journal of the Learning Sciences, 12*(3), 307–359.

Barry, M., Small, M., Avard-Spinney, A., & Burnard Wheadon, L. (2000). *Mathematical modeling: Book 1*. Scarborough: Nelson Thompson.

Barton, B. (2008). *The language of mathematics: Telling mathematical tales*. New York: Springer.

Barton, B., Fairhall, U., & Trinick, T. (1998). *Tikanga reo tatai*: Issues in the development of a Maori mathematics register. *For the Learning of Mathematics, 18*(1), 3–9.

Barwell, R. (2004). *Teaching learners of English as an additional language: A review of official guidance*, NALDIC Working Paper 6. Watford: National Association for Language Development in the Curriculum Publications Group.

Barwell, R. (2005a). Critical issues for language and content in mainstream classrooms: Introduction. *Linguistics and Education, 16*(2), 143–150.

Barwell, R. (ed.) (2005b). Language and content in mainstream classrooms. Special issue of *Linguistics and Education, 16*(2), 143–252.

Barwell, R. (2007). Semiotic resources for doing and learning mathematics. *For the Learning of Mathematics, 27*(1), 31–32.

Barwell, R. (2009). Researchers' descriptions and the construction of mathematical thinking. *Educational Studies in Mathematics, 72*(2), 255–269.

Battiste, M. (Ed.). (2000). *Reclaiming indigenous voice and vision*. Vancouver: University of British Columbia Press.

Baturo, A., Norton, S., & Cooper, T. (2004). The mathematics of indigenous card games: Implications for mathematics teaching and learning. In I. Putt, R. Faragher, & M. McLean (Eds.), *Mathematics education for the third millennium: Towards 2010 (Proceedings of the 27th annual MERGA conference)* (pp. 87–94). Sydney: Mathematics Education Research Group of Australasia.

Baugh, J. (1994). New and prevailing misconceptions of African American English for logic and mathematics. In E. Hollins, J. King, & W. Hayman (Eds.), *Teaching diverse populations: Formulating a knowledge base* (pp. 191–206). Albany: State University of New York Press.

Baxter, J., & Williams, S. (2010). Social and analytic scaffolding in middle school mathematics: Managing the dilemma of telling. *Journal of Mathematics Teacher Education, 13*(1), 7–26.

Benton, R. (1978). Problems and prospects for indigenous languages and bilingual education in New Zealand and Oceania. In B. Spolsky & R. Cooper (Eds.), *Case studies in bilingual education* (pp. 126–166). Rowley: Newbury House Publishers.

Bernstein, B. (1982). Codes, modalities and the process of cultural reproduction: A model. In M. Apple (Ed.), *Cultural and economic reproduction in education* (pp. 304–355). London: Routledge and Kegan Paul.

Bernstein, B. (1990). *The structuring of pedagogic discourse*. London: Routledge.

Bernstein, B. (2000, rev'd edn). *Pedagogy, symbolic control and identity: Theory, research and critique*. Lanham: Rowman and Littlefield.

Bills, L. (2000). Politeness in teacher–student dialogue in mathematics: A socio-linguistic analysis. *For the Learning of Mathematics, 20*(2), 40–47.

Bishop, A. (1988). *Mathematical enculturation: A cultural perspective on mathematics education*. Dordrecht: Kluwer Academic Publishers.

Blachowicz, C., & Fisher, P. (2000). Vocabulary instruction. In M. Kamil, P. Mosenthal, P. Pearson, & R. Barr (Eds.), *Handbook of reading research* (Vol. 3, pp. 503–523). Mahwah: Lawrence Erlbaum Associates.

Boaler, J. (1997). *Experiencing school mathematics: Teaching styles, sex and setting*. Buckingham: Open University Press.

Boaler, J. (2003). When learning no longer matters: Standardized testing and the creation of inequality. *Phi Delta Kappan, 84*(7), 502–506.

Boaler, J. (2006a). How a detracked mathematics approach promoted respect, responsibility, and high achievement. *Theory into Practice, 45*(1), 40–46.

Boaler, J. (2006b). Urban success: A multidimensional mathematics approach with equitable outcomes. *Phi Delta Kappan, 87*(5), 364–369.

Boaler, J. (2008). Promoting "relational equity" and high mathematics achievement through an innovative mixed-ability approach. *British Educational Research Journal, 34*(2), 167–194.

Boaler, J., & Staples, M. (2008). Creating mathematical futures through an equitable teaching approach: The case of Railside school. *Teachers College Record, 110*(3), 608–645.

Boaler, J., Cobb, P., Gresalfi, M., Horn, I., Hodge, L., & Staples, M. (2006). *Equity and identity in mathematics classrooms: Students' learning opportunities, knowledge orientation, persistence, and "future selves"*. Paper presented at the annual meeting of the American Educational Research Association, San Francisco.

Borman, K. (2005). *Meaningful urban education reform: Confronting the learning crisis in mathematics and science*. Albany: State University of New York Press.

Bourdieu, P. (1986). The forms of capital. In J. Richardson (Ed.), *Handbook of theory and research for the sociology of education* (pp. 241–258). New York: Greenwood Press.

Bourdieu, P. (1990). *The logic of practice*. Cambridge: Polity Press.

Bourdieu, P. (1991). *Language and symbolic power*. Cambridge: Harvard University Press.

Bourdieu, P. (1998). *Practical reason: On the theory of action*. Cambridge: Polity Press.

Bourdieu, P., Passeron, J., & de Saint Martin, M. (1994a). *Academic discourse: Linguistic misunderstanding and professorial power*. Cambridge: Polity Press.

Bourdieu, P., Passeron, J., & de saint Martin, M. (1994b). Students and the language of teaching. In P. Bourdieu, J. Passeron, & M. de saint Martin (Eds.), *Academic discourse: Linguistic misunderstanding and professorial power* (pp. 35–79). Cambridge: Polity Press.

Bourne, J. (2003). Vertical discourse: The role of the teacher in the transmission and acquisition of decontextualised language. *European Educational Research Journal, 2*(4), 496–521.

Braddock, J., & Dawkins, M. (1993). Ability grouping, aspirations, and attainments: Evidence from the National Educational Longitudinal Study of 1988. *Journal of Negro Education, 62*(3), 324–336.

Brenner, M. (1998). Adding cognition to the formula for culturally relevant instruction in mathematics. *Anthropology and Education Quarterly, 29*(2), 214–244.

Brown, A., & Palincsar, A. (1989). Guided, cooperative learning and individual knowledge acquisition. In L. Resnick (Ed.), *Knowing, learning, and instruction: Essays in honor of Robert Glaser* (pp. 393–452). Hillsdale: Lawrence Erlbaum Associates.

Bruner, J. (1986). *Actual minds, possible worlds*. Cambridge: Harvard University Press.

Buckley, L. (2010). Unfulfilled hopes in education for equity: Redesigning the mathematics curriculum in a US high school. *Journal of Curriculum Studies, 42*(1), 51–78.

Burton, L., & Morgan, C. (2000). Mathematicians writing. *Journal for Research in Mathematics Education, 31*(4), 429–453.

Cajete, G. (1994). *Look to the mountain: An ecology of indigenous education*. Durango: Kivaki Press.

Calvino, I. (1988/1992). *Six memos for the next millennium*. London: Jonathan Cape.

Cavanagh, S. (2005). Math: The not-so-universal language. *Education Week, 241*(42), 1–22.

Cazden, C. (2001). *Classroom discourse: The language of teaching and learning* (2nd ed.). Portsmouth: Heinemann.

Cazden, C., & Mehan, H. (1992). Principles from sociology and anthropology: Context, code, classroom, and culture. In M. Reynolds (Ed.), *Knowledge base for the beginning teacher* (pp. 47–57). Toronto: Pergamon Press.

CCSS. (2010). *Common core state standards for mathematics*. National Governors' Association Center for Best Practices and Council of Chief State School Officers. (http://www.corestandards.org/the-standards)

Chamberlin, M. (2005). Teachers' discussions of students' thinking: Meeting the challenge of attending to students' thinking. *Journal of Mathematics Teacher Education, 8*(2), 141–170.

Chapman, A. (1995). Intertextuality in school mathematics: The case of functions. *Linguistics and Education, 7*(3), 243–262.

Chapman, A. (2003). *Language practices in school mathematics: A social semiotic approach.* Lewiston: Edwin Mellen Press.

Chazan, D. (2000). *Beyond formulas in mathematics and teaching: Dynamics of the high school algebra classroom.* New York: Teachers College Press.

Chazan, D., & Ball, D. (1999). Beyond being told not to tell. *For the Learning of Mathematics, 19*(2), 2–10.

Chizhik, A. (2001). Equity and status in group collaboration: Learning through explanations depends on task characteristics. *Social Psychology of Education, 5*(2), 179–200.

Chizhik, A., Alexander, M., Chizhik, E., & Goodman, J. (2003). The rise and fall of power and prestige orders: Influence of task structure. *Social Psychology Quarterly, 66*(3), 303–317.

Chouliaraki, L., & Fairclough, N. (1999). *Discourse in late modernity: Rethinking critical discourse analysis.* Edinburgh: Edinburgh University Press.

Christie, F. (2002). *Classroom discourse analysis: A functional perspective.* London: Continuum.

Civil, M. (2006). Working towards equity in mathematics education: A focus on learners, teachers and parents. In S. Alatorre, J. Cortina, M. Sáiz, & A. Méndez (Eds.), *Proceedings of the 28th annual meeting of the North American Chapter of the International Group of the Psychology of Mathematics Education* (Vol. 1, pp. 30–50). Mérida: Universidad Pedagógica Nacional.

Civil, M., & Planas, N. (2004). Participation in the mathematics classroom: Does every student have a voice? *For the Learning of Mathematics, 24*(1), 7–12.

Cobb, P. (2000). The importance of a situated view of learning to the design of research and instruction. In J. Boaler (Ed.), *Multiple perspectives on mathematics teaching and learning* (pp. 45–82). Westport: Ablex Corporation.

Cobb, P., & Hodge, L. (2002). A relational perspective on issues of cultural diversity and equity as they play out in the mathematics classroom. *Mathematical Thinking and Learning, 4*(2/3), 249–284.

Cobb, P., Stephan, M., McClain, K., & Gravemeijer, K. (2001). Participating in classroom mathematical practices. *The Journal of the Learning Sciences, 10*(1–2), 113–163.

Cobb, P., Wood, T., & Yackel, E. (1991). A constructivist approach to second grade mathematics. In E. von Glasersfeld (Ed.), *Radical constructivism in mathematics education* (pp. 157–176). Dordrecht: Kluwer Academic Publishers.

Cobb, P., Wood, T., Yackel, E., & McNeal, B. (1992). Characteristics of classroom mathematics traditions: An interactional analysis. *American Educational Research Journal, 29*(3), 573–604.

Cochran-Smith, M., & Lytle, S. (1999). Relationships of knowledge and practice: Teacher learning in communities. *Review of Research in Education, 24*(1), 249–305.

Cohen, D. (1995). What is the system in systemic reform? *Educational Researcher, 24*(9), 11–17, 31.

Cohen, E. (1994). Restructuring the classroom: Conditions for productive small groups. *Review of Educational Research, 64*(1), 1–35.

Cohen, E., & Lotan, R. (Eds.). (1997). *Working for equity in heterogeneous classrooms: Sociological theory in practice.* New York: Teachers College Press.

Cohen, E., Lotan, R., Scarloss, B., & Arellano, A. (1999). Complex instruction: Equity in cooperative learning classrooms. *Theory into Practice, 38*(2), 80–86.

Cole, M. (1996). *Cultural psychology: A once and future discipline.* Cambridge: Belknap/Harvard University Press.

Cooper, B., & Dunne, M. (1999). *Assessing children's mathematical knowledge: Social class, sex and problem solving.* Buckingham: Open University Press.

Cotton, T., & Gates, P. (1996). Why the psychological must consider the social in promoting equity and social justice in mathematics education. In L. Puig & A. Gutiérrez (Eds.), *Proceedings of the 20th Conference of the International Group for the Psychology of Mathematics Education* (Vol. 2, pp. 249–256). València: Universitat de València.

Crespo, S. (2006). Elementary teacher talk in mathematics study groups. *Educational Studies in Mathematics, 63*(1), 29–56.

Crockett, M. (2002). Inquiry as professional development: Creating dilemmas through teachers' work. *Teaching and Teacher Education, 18*(5), 609–624.

Cross, M., Shalem, Y., Backhouse, J., Adam, F., & Baloyi, H. (2010). "Wits gives you the edge": How students negotiate the pressures of undergraduate study. In *Higher education monitor: Access and throughput in South African Higher Education – Three case studies* (pp. 54–94). Pretoria: Council for Higher Education.

Crowhurst, M. (1994). *Language and learning across the curriculum*. Scarborough: Allyn and Bacon.

Cummins, J. (2000). *Language, power and pedagogy: Bilingual children in the crossfire*. Clevedon: Multilingual Matters.

D'Ambrosio, U. (1985). Ethnomathematics and its place in the history and pedagogy of mathematics. *For the Learning of Mathematics, 5*(1), 44–48.

D'Ambrosio, U. (1994). Cultural framing of mathematics teaching and learning. In R. Biehler, R. Scholz, R. Sträβer, & B. Winkelmann (Eds.), *Didactics of mathematics as a scientific discipline* (pp. 443–455). Dordrecht: Kluwer Academic Publishers.

D'Ambrosio, U. (1997). Where does ethnomathematics stand nowadays? *For the Learning of Mathematics, 17*(2), 13–17.

D'Ambrosio, U. (2006). *Ethnomathematics: Link between traditions and modernity*. Rotterdam: Sense Publishers.

Damon, W. (1984). Peer education: The untapped potential. *Journal of Applied Developmental Psychology, 5*(4), 331–343.

Darling-Hammond, L. (1994). Performance-based assessment and educational equity. *Harvard Educational Review, 64*(1), 5–30.

Darling-Hammond, L. (1995). Inequality and access to knowledge. In J. Banks & C. McGee Banks (Eds.), *Handbook of research on multicultural education* (pp. 465–483). New York: Macmillan.

Darling-Hammond, L. (1997). *The right to learn: A blueprint for creating schools that work*. San Francisco: Jossey-Bass Publishers.

Davies, B., & Harré, R. (1990). Positioning: The discursive production of selves. *Journal for the Theory of Social Behavior, 20*(1), 43–63.

Davies, B., & Harré, R. (1999). Positioning and personhood. In R. Harré & L. van Langenhove (Eds.), *Positioning theory: Moral contexts of intentional action* (pp. 32–51). Oxford: Blackwell.

de Abreu, G. (1999). Learning mathematics in and outside of school: Two views on situated learning. In J. Bliss, R. Säljö, & P. Light (Eds.), *Learning sites: Social and technological resources for learning* (pp. 17–31). Oxford: Elsevier.

de Abreu, G., & Cline, T. (2007). Social valorization of mathematical practices: The implications for learners in multicultural schools. In N. Nasir & P. Cobb (Eds.), *Improving access to mathematics: Diversity and equity in the classroom* (pp. 118–131). New York: Teachers College Press.

De Avila, E. (1988). Bilingualism, cognitive function and language minority group membership. In R. Cocking & J. Mestre (Eds.), *Linguistic and cultural influences on learning mathematics* (pp. 101–121). Hillsdale: Lawrence Erlbaum Associates.

de Freitas, E., & Zolkower, B. (2009). Using social semiotics to prepare mathematics teachers to teach for social justice. *Journal of Mathematics Teacher Education, 12*(3), 187–203.

de Freitas, E., & Zolkower, B. (2011). Developing teacher capacity to explore non-routine problems through a focus on the social semiotics of mathematics classroom discourse. *Research in Mathematics Education, 13*(3), 229–248.

de Haan, M., & Elbers, E. (2005). Reshaping diversity in a local classroom: Communication and identity issues in multicultural schools in the Netherlands. *Language and Communication, 25*(3), 315–333.

DeBose, C. (2007). The Ebonics phenomenon, language planning and the hegemony of standard English. In H. Alim & J. Baugh (Eds.), *Talkin Black Talk: Language, education, and social change* (pp. 30–42). New York: Teachers College Press.

Delpit, L. (1995). *Other people's children: Cultural conflict in the classroom*. New York: New Press.
DfES. (2000). *Mathematical vocabulary*. London: Department for Education and Skills.
DfES. (2002). *Access and engagement in mathematics: Teaching pupils for whom English is an additional language*. London: Department for Education and Skills.
DfES. (2006). *Primary framework for literacy and mathematics*. London: Department for Education and Skills.
DIAND. (1969). *Statement of the Government of Canada on Indian Policy*. Ottawa: Department of Indian Affairs and Northern Development.
Doolittle, E. (2006). Mathematics as medicine. In P. Liljedahl (Ed.), *Proceedings of the 2006 annual meeting of the Canadian Mathematics Education Study Group* (pp. 17–25). Burnaby: Simon Fraser University.
Dowling, P. (1991). A touch of class: ability, social class and intertext in SMP 11-16. In D. Pimm & E. Love (Eds.), *Teaching and learning school mathematics* (pp. 137–152). London: Hodder and Stoughton.
Dowling, P. (1998). *The sociology of mathematics education: Mathematical myths/pedagogic texts*. London: Falmer Press.
Eckert, P. (1989). *Jocks and burnouts: Social categories and identity in the high school*. New York: Teachers College Press.
Eco, U. (1994). *Six walks in the fictional woods*. Cambridge: Harvard University Press.
Education Trust. (2010). *United States is average in performance, but leads the world in inequity*. (http://www.edtrust.org/print/2125)
Edwards, D. (1997). *Discourse and cognition*. London: Sage.
Edwards, D. (2006). Discourse, cognition and social practices: The rich surface of language and social interaction. *Discourse Studies, 8*(1), 41–49.
Edwards, D., & Mercer, N. (1987). *Common knowledge: The development of understanding in the classroom*. New York: Methuen.
Edwards, D., & Potter, J. (1992). *Discursive psychology*. London: Sage.
El Barrio–Hunter College PDS Partnership Writing Collective. (2009). On the unique relationship between teacher research and commercial mathematics curriculum development. In J. Remillard, B. Herbel-Eisenmann, & G. Lloyd (Eds.), *Mathematics teachers at work: Connecting curriculum materials and classroom instruction* (pp. 118–133). New York: Routledge.
Engle, R., & Conant, F. (2002). Guiding principles for fostering productive disciplinary engagement: Explaining an emergent argument in a community of learners classroom. *Cognition and Instruction, 20*(4), 399–483.
Erickson, F. (1982). Classroom discourse as improvisation: Relationships between academic task structure and social participation structure in lessons. In L. Wilkinson (Ed.), *Communicating in the classroom* (pp. 153–182). New York: Academic.
Ervynck, G. (1992). Mathematics as a foreign language. In W. Geeslin & K. Graham (Eds.), *Proceedings of the sixteenth conference of the International Group for the Psychology of Mathematics Education* (Vol. 3, pp. 217–233). Durham: University of New Hampshire.
Esmonde, I. (2006). *"How are we supposed to, like, learn it, if none of us know?": Opportunities to learn and equity in mathematics cooperative learning structures*. Unpublished doctoral dissertation. University of California, Berkeley.
Esmonde, I. (2009). Ideas and identities: Supporting equity in cooperative mathematics learning. *Review of Educational Research, 79*(2), 1008–1043.
Evans, J. (2000). *Adults' mathematical thinking and emotions: A study of numerate practices*. London: RoutledgeFalmer.
Ezeife, A. (2002). Mathematics and culture nexus: The interactions of culture and mathematics in an aboriginal classroom. *International Education Journal, 3*(3), 176–187.
Fairclough, N. (1992). *Critical language awareness*. London: Longman.
Fairclough, N. (1995). *Critical discourse analysis: The critical study of language*. Harlow: Longman.

Fairclough, N. (2001). *Language and power* (2nd ed.). New York: Longman.

Fairclough, N. (2003). *Analysing discourse: Textual analysis for social research*. London: Routledge.

Fang, Z., & Schleppegrell, M. (2008). *Reading in secondary content areas: A language-based pedagogy*. Ann Arbor: University of Michigan Press.

Fantuzzo, J., King, J., & Heller, J. (1992). Effects of reciprocal peer tutoring on mathematics and school adjustment: A component analysis. *Journal of Educational Psychology, 84*(3), 331–339.

Farivar, S., & Webb, N. (1994). Helping and getting help: Essential skills for effective group problem solving. *Arithmetic Teacher, 41*(9), 521–525.

Fauvel, J. (1987). *The renaissance of mathematical sciences in Britain (unit 6)*. Milton Keynes: The Open University.

Fernandez, C., & Yoshida, M. (2004). *Lesson study: A Japanese approach to improving mathematics teaching and learning*. Mahwah: Lawrence Erlbaum Associates.

Finegan, E., & Besnier, N. (1989). *Language: Its structure and use*. New York: Harcourt Brace Jovanovich.

Forman, E. (1996). Learning mathematics as participation in classroom practice: Implications of sociocultural theory for educational reform. In L. Steffe, P. Nesher, P. Cobb, G. Goldin, & B. Greer (Eds.), *Theories of mathematical learning* (pp. 115–130). Mahwah: Lawrence Erlbaum Associates.

Forman, E. (2003). A sociocultural approach to mathematics reform: Speaking, inscribing, and doing mathematics within a community of practice. In J. Kilpatrick, W. Martin, & D. Schifter (Eds.), *A research companion to principles and standards for school mathematics* (pp. 333–352). Reston: National Council of Teachers of Mathematics.

Forman, E., & Ansell, E. (2002). Orchestrating the multiple voices and inscriptions of a mathematics classroom. *The Journal of the Learning Sciences, 11*(2–3), 251–274.

Forman, E., Larreamendy-Joerns, J., Stein, M., & Brown, C. (1998). "You're going to want to find out which and prove it": Collective argumentation in a mathematics classroom. *Learning and Instruction, 8*(6), 527–548.

Foucault, M. (1972). *The archaeology of knowledge*. London: Routledge.

Foucault, M. (1975/1984). 'The means of correct training' and 'Panopticism'. In Rainbow, P. (Ed.), *The Foucault reader* (pp. 188–213). New York: Pantheon Books.

Franke, M., & Kazemi, E. (2001). Teaching as learning within a community of practice: Characterizing generative growth. In T. Wood, B. Nelson, & J. Warfield (Eds.), *Beyond classical pedagogy* (pp. 47–74). Mahwah: Lawrence Erlbaum Associates.

Franke, M., Kazemi, E., & Battey, D. (2007). Understanding teaching and classroom practice in mathematics. In F. Lester (Ed.), *Second handbook of research on mathematics teaching and learning* (pp. 225–256). Charlotte: Information Age Publishing/National Council of Teachers of Mathematics.

Frankenstein, M. (1998). Reading the world and math: Goals for a critical mathematical literacy curriculum. In E. Lee, D. Menkart, & M. Okazawa-Rey (Eds.), *Beyond heroes and holidays: A practical guide to K–12 anti-racist, multicultural education and staff development* (pp. 306–313). Washington, DC: Network of Educators of the Americas.

Freebody, P., Ludwig, C., & Gunn, S. (1995). *Everyday literacy practice in and out of schools in low socioeconomic urban communities*. Brisbane: Griffith University, Centre for Literacy Education Research.

Fuhrman, S. (2004). Introduction. In S. Fuhrman & R. Elmore (Eds.), *Redesigning accountability systems for education* (pp. 3–14). New York: Teachers College Press.

Gamoran, A. (1987). The stratification of high school learning opportunities. *Sociology of Education, 60*(3), 135–155.

García, E., & González, R. (1995). Issues in systemic reform for culturally and linguistically diverse students. *Teachers College Record, 96*(3), 418–431.

Gay, G. (2000). *Culturally responsive teaching: Theory, research, and practice*. New York: Teachers College Press.

Gee, J. (1996). *Social linguistics and literacies: Ideology in discourses* (2nd ed.). London: Taylor & Francis.

Gee, J. (1999). *An introduction to discourse analysis: Theory and method.* London: Routledge.
Gee, J. (2008). A sociocultural perspective on opportunity to learn'. In P. Moss, D. Pullin, J. Gee, E. Haertel, & L. Young (Eds.), *Assessment, equity and opportunity to learn* (pp. 76–108). New York: Cambridge University Press.
Gerdes, P. (1988). On possible uses of traditional Angolan sand drawings in the mathematics classroom. *Educational Studies in Mathematics, 19*(1), 3–22.
Gerdes, P. (1997). Survey of current work on ethnomathematics. In A. Powell & M. Frankenstein (Eds.), *Ethnomathematics: Challenging eurocentrism in mathematics education* (pp. 331–371). Albany: State University of New York Press.
Gerofsky, S. (2004). *A man left Albuquerque heading east: Word problems as genre in mathematics education.* New York: Peter Lang.
Goffman, E. (1981). *Forms of talk.* Philadelphia: University of Pennsylvania Press.
Goldberg, S. (2010). Tony Judt: The captivating wit and intellect of my friend and teacher. *The Guardian*, August 7th. (www.guardian.co.uk/commentisfree/2010/aug/07/tony-judt-eulogy-saul-goldberg)
González, N. (1995). Processual approaches to multicultural education. *Journal of Applied Behavioral Science, 31*(2), 234–244.
González, G. (2009). *Mathematical tasks and the collective memory: How do teachers manage students' prior knowledge when teaching geometry with problems?* Unpublished Ph.D. dissertation. University of Michigan, Ann Arbor.
González, G. (2011). *Who* does *what*? A linguistic approach to analyzing teachers' reactions to videos. *ZDM The International Journal on Mathematics Education, 43*(1), 65–80.
González, N., Andrade, R., Civil, M., & Moll, L. (2001). Bridging funds of distributed knowledge: Creating zones of practices in mathematics. *Journal of Education for Students Placed at Risk, 6*(1–2), 115–132.
Goody, J. (1977). *The domestication of the savage mind.* Cambridge: Cambridge University Press.
Grant, M., & McGraw, R. (2006). Collaborating to investigate and improve classroom mathematics discourse. In L. van Zoest (Ed.), *Teachers engaged in research: Inquiry into mathematics classrooms, grades 9–12* (pp. 231–251). Greenwich: Information Age Publishing.
Granville, S., Janks, H., Mphahlele, M., Reed, Y., Joseph, M., Ramani, E., & Watson, P. (1998). English with or without g(u)ilt: A position paper on language in education policy for south Africa. *Language and Education, 12*(4), 254–272.
Gray, B. (1999). Literacy, numeracy, attendance and health. *Unicorn, 25*(3), 17–18.
Greeno, J., & Gresalfi, M. (2008). Opportunities to learn in practice and identity. In P. Moss, D. Pullin, J. Gee, E. Haertel, & L. Young (Eds.), *Assessment, equity and opportunity to learn* (pp. 170–199). New York: Cambridge University Press.
Gronewold, P. (2009). "Math is about thinking": From increased participation to conceptual talk. In B. Herbel-Eisenmann & M. Cirillo (Eds.), *Promoting purposeful discourse: Teacher research in mathematics classrooms* (pp. 45–56). Reston: National Council of Teachers of Mathematics.
Gutiérrez, K., & Rogoff, B. (2003). Cultural ways of learning: Individual traits or repertoires of practice? *Educational Researcher, 32*(5), 19–25.
Gutiérrez, K., Baquedano-Lopez, P., & Alvarez, H. (2001). Literacy as hybridity: Moving beyond bilingualism in urban classrooms. In M. de la Luz Reyes & J. Halcón (Eds.), *The best for our children: Critical perspectives on literacy for Latino students* (pp. 122–141). New York: Teachers College Press.
Gutiérrez, K., Rymes, B., & Larson, J. (1995). Script, counterscript, and underlife in the classroom: James Brown versus *Brown v. Board of Education. Harvard Educational Review, 65*(3), 445–471.
Gutiérrez, R. (1996). Practices, beliefs and cultures of high school mathematics departments: Understanding their influence on student advancement. *Journal of Curriculum Studies, 28*(5), 495–529.

Gutiérrez, R. (1998). Departments as contexts for understanding and reforming secondary teachers' work: Continuing the dialogue. *Journal of Curriculum Studies, 30*(1), 95–104.

Gutiérrez, R. (1999). Advancing urban Latina/o youth in mathematics: Lessons from an effective high school mathematics department. *The Urban Review, 31*(3), 263–281.

Gutiérrez, R. (2002a). Enabling the practice of mathematics teachers in context: Toward a new equity research agenda. *Mathematical Thinking and Learning, 4*(2/3), 145–187.

Gutiérrez, R. (2002b). Beyond essentialism: The complexity of language in teaching mathematics to Latina/o students. *American Educational Research Journal, 39*(4), 1047–1088.

Gutiérrez, R. (2004). *"So that's what it means to teach urban Latina/o students quality mathematics": A community of practice model of teacher education.* Paper presented at the annual meeting of the American Educational Research Association, New Orleans.

Gutiérrez, R. (2005). *Organizing for advancement: Can teacher communities foster equity?.* Plenary address, Annual meeting of the National Council of Teachers of Mathematics, research pre-session, Los Angeles.

Gutiérrez, R. (2007a). (Re)defining equity: The importance of a critical perspective. In N. Nasir & P. Cobb (Eds.), *Improving access to mathematics: Diversity and equity in the classroom* (pp. 37–50). New York: Teachers College Press.

Gutiérrez, R. (2007b). Context matters: Equity, success, and the future of mathematics education. In T. Lamberg & L. Wiest (Eds.), *Proceedings of the 29th annual meeting of the North American Chapter of the International Group for the Psychology of Mathematics Education* (pp. 1–18). Stateline: University of Nevada, Reno.

Gutiérrez, R. (2008). A "gap-gazing" fetish in mathematics education? Problematizing research on the achievement gap. *Journal for Research in Mathematics Education, 39*(4), 356–364.

Gutiérrez, R. (2009). Embracing the inherent tensions in teaching mathematics from an equity stance. *Democracy and Education, 18*(3), 9–16.

Gutiérrez, R. (2010). The sociopolitical turn in mathematics education. *Journal for Research in Mathematics Education.* 41(0), 1–32. (www.nctm.org/publications/toc.aspx?jrnl=JRME&mn=6&y=2010)

Gutiérrez, R. (in press). Stand and deliver: The challenge of language in the study of mathematics. In Garcia, M. & Valdivia, A. (Eds.), *Mapping Latina/Latino studies for the twenty-first century.* New York: Peter Lang.

Gutiérrez, R., & Dixon-Román, E. (2011). Beyond gap gazing: How can thinking about education comprehensively help us (re)envision mathematics education? In B. Atweh, M. Graven, W. Secada, & P. Valero (Eds.), *Mapping equity and quality in mathematics education* (pp. 21–34). New York: Springer.

Gutiérrez, R., & Morales, H. (2002). Teacher community, socialization and biography in reforming mathematics. In V. Lee & A. Bryk (Eds.), *Reforming Chicago's high schools: Research perspectives on school and system level change* (pp. 223–249). Chicago: Consortium on Chicago School Research.

Gutstein, E. (2003). Teaching and learning mathematics for social justice in an urban, Latino school. *Journal for Research in Mathematics Education, 34*(1), 37–73.

Gutstein, E. (2006). *Reading and writing the world with mathematics: Toward a pedagogy for social justice.* New York: Routledge.

Hakuta, K., & McLaughlin, B. (1996). Bilingualism and second language learning: Seven tensions that define the research. In D. Berliner & R. Calfee (Eds.), *Handbook of educational psychology* (pp. 603–621). New York: Macmillan.

Halliday, M. (1975a). Some aspects of sociolinguistics. In *Interactions between linguistics and mathematical education* (pp. 64–73). Copenhagen: Royal Danish School of Educational Studies.

Halliday, M. (1975b). *Learning How to mean: Explorations in the development of language.* London: Edward Arnold.

Halliday, M. (1978). Sociolinguistic aspects of mathematics education. In M. Halliday (Ed.), *Language as social semiotic: The social interpretation of language and meaning* (pp. 194–204). London: Edward Arnold.

Halliday, M., & Matthiessen, C. (2004). *An introduction to functional grammar* (3rd ed.). London: Edward Arnold.

Hallinan, M. (1994). Tracking: From theory to practice. *Sociology of Education, 67*(2), 79–84.

Hand, V. (2010). The co-construction of opposition in a low-track mathematics classroom. *American Educational Research Journal, 47*(1), 97–132.

Haney, W. (2000). The myth of the Texas miracle in education. *Education Policy Analysis Archives 8*(41). (http://epaa.asu.edu)

Hannula, M. (2002). Attitude toward mathematics: Emotions, expectations, and values. *Educational Studies in Mathematics, 49*(1), 25–46.

Harré, R., & van Langenhove, L. (Eds.). (1999). *Positioning theory: Moral contexts of intentional action*. Oxford: Blackwell.

Harris, P. (1991). *Mathematics in a cultural context: Aboriginal perspectives on space, time and money*. Geelong: Deakin University Press.

Hayes, D., Mills, M., Christie, P., & Lingard, B. (2006). *Teaching and schooling making a difference: Productive pedagogies, assessment and performance*. Crows Nest: Allen Unwin.

Heath, S. (1982). Questioning at home and at school: A comparative study. In G. Spindler (Ed.), *Doing the ethnography of schooling* (pp. 102–131). New York: Holt, Rinehart and Winston.

Heath, S. (1983). *Ways with words: Language, life and work in communities and classrooms*. Cambridge: Cambridge University Press.

Heath, S. (1986). Sociocultural contexts of language development. In California, Office of *Bilingual Bicultural Education, Beyond Language: Social and Cultural Factors in Schooling Language Minority Students* (pp. 143–186), Developed by the Bilingual Education Office, California State Department of Education, Sacramento. Los Angeles: California State University, Los Angeles, Evaluation, Dissemination and Assessment Center.

Hedges, L., & Nowell, A. (1999). Changes in the Black–White gap in achievement test scores. *Sociology of Education, 72*(2), 111–135.

Herbel-Eisenmann, B. (2007). From intended curriculum to written curriculum: Examining the "voice" of a mathematics textbook. *Journal for Research in Mathematics Education, 38*(4), 344–369.

Herbel-Eisenmann, B. (2009). Negotiating the "presence *of* the text": How might teachers' language choices influence the positioning of the textbook? In J. Remillard, B. Herbel-Eisenmann, & G. Lloyd (Eds.), *Mathematics teachers at work: Connecting curriculum materials and classroom instruction* (pp. 134–151). New York: Routledge.

Herbel-Eisenmann, B. (2010). Beyond tacit language choice to purposeful discourse practices. In L. Knott (Ed.), *The role of mathematics discourse in producing leaders of discourse* (pp. 173–198). Charlotte: Information Age Publishing.

Herbel-Eisenmann, B., & Breyfogle, M. (2005). Questioning our *patterns* of questioning. *Mathematics Teaching in the Middle School, 10*(9), 484–489.

Herbel-Eisenmann, B., & Cirillo, M. (Eds.). (2009). *Promoting purposeful discourse: Teacher research in mathematics classrooms*. Reston: National Council of Teachers of Mathematics.

Herbel-Eisenmann, B., & Otten, S. (2011). Mapping mathematics in classroom discourse. *Journal for Research in Mathematics Education, 42*(5), 451–485.

Herbel-Eisenmann, B., & Schleppegrell, M. (2008). "What question would I be asking myself in my head?": Helping all students reason mathematically. In M. Ellis (Ed.), *Mathematics for every student: Responding to diversity, grades 6–8* (pp. 23–37). Reston: National Council of Teachers of Mathematics.

Herbel-Eisenmann, B., & Wagner, D. (2005). In the middle of nowhere: How a textbook can position the mathematics learner. In H. Chick & J. Vincent (Eds.), *Proceedings of the 29th conference of the International Group for the Psychology of Mathematics Education* (Vol. 3, pp. 121–128). Melbourne: University of Melbourne.

Herbel-Eisenmann, B., & Wagner, D. (2007). A framework for uncovering the way a textbook may position the mathematics learner. *For the Learning of Mathematics, 27*(2), 8–14.

Herbel-Eisenmann, B., & Wagner, D. (2010). Appraising lexical bundles in mathematics classroom discourse: Obligation and choice. *Educational Studies in Mathematics, 75*(1), 43–63.

Herbel-Eisenmann, B., Cirillo, M., & Skowronski, K. (2009a). Why discourse deserves our attention! In A. Flores (Ed.), *Mathematics for every student: Responding to diversity, grades 9–12* (pp. 103–115). Reston: National Council of Teachers of Mathematics.

Herbel-Eisenmann, B., Drake, C., & Cirillo, M. (2009b). "Muddying the clear waters": Teachers' take-up of the linguistic idea of revoicing. *Teaching and Teacher Education, 25*(2), 268–277.

Herbel-Eisenmann, B., Wagner, D., & Cortes, V. (2010). Lexical bundle analysis in mathematics classroom discourse: The significance of stance. *Educational Studies in Mathematics, 75*(1), 23–42.

Hilliard, A. (2003). No mystery: Closing the achievement gap between Africans and excellence. In T. Perry, C. Steele, & A. Hilliard (Eds.), *Young, gifted and black: Promoting high achievement among African American students* (pp. 131–165). Boston: Beacon.

Hodge, R., & Kress, G. (1993). *Language as ideology* (2nd ed.). London: Routledge.

Hornberger, N. (1988). *Bilingual education and language maintenance: A Southern Peruvian Quechua case*. Dordrecht: Foris Publications.

Howard, T. (2008). Who really cares? The disenfranchisement of African American males in preK–12 schools: A critical race theory perspective. *Teachers College Record, 110*(5), 954–985.

Howie, S. (2003). Language and other background factors affecting secondary pupils' performance in mathematics in South Africa. *African Journal of Research in Mathematics Science and Technology Education, 7*(1), 1–20.

Howie, S. (2004). A national assessment in mathematics within an international comparative assessment. *Perspectives in Education, 22*(2), 149–162.

HREOC. (2000). *Emerging themes: National inquiry into rural and remote education*. Sydney: Human Rights and Equal Opportunity Commission.

Huang, J., & Normandia, B. (2008). Comprehending and solving word problems in mathematics: Beyond key words. In Fang, Z., & Schleppegrell, M. *Reading in secondary content areas: A language-based pedagogy* (pp. 64–83). Ann Arbor: University of Michigan Press.

INAC. (2002). *Our children – Keepers of our sacred knowledge* (The final report of the minister's national working group on education). Ottawa: Indian and Northern Affairs Canada. (http://www.ainc-inac.gc.ca/ps/edu/finre/bac_e.html)

Indian Chiefs of Alberta. (1970). *Citizens plus (red paper)*. Edmonton: Indian Association of Alberta.

Ingersoll, R. (1999). The problem of underqualified teachers in American secondary schools. *Educational Researcher, 28*(2), 26–37.

Ingersoll, R., & Gruber, K. (1996). *Out-of-field teaching and educational equality*. Washington, DC: National Center for Educational Statistics.

Joseph, G. (1991). *The crest of the peacock: Non-European roots of mathematics*. London: Tauris.

Kahn, M. (2001). Changing science and mathematics achievement: Reflections on policy and planning. *Perspectives in Education, 19*(3), 169–176.

Khisty, L. (1995). Making inequality: Issues of language and meaning in mathematics teaching with Hispanic students. In W. Secada, E. Fennema, & L. Adajian (Eds.), *New directions for equity in mathematics education* (pp. 279–297). New York: Cambridge University Press.

Khisty, L., & Chval, K. (2002). Pedagogic discourse and equity in mathematics: When teachers' talk matters. *Mathematics Education Research Journal, 14*(3), 154–168.

Khisty, L., & Viego, G. (1999). Challenging conventional wisdom: A case study. In L. Ortiz-Franco, N. Hernandez, & Y. de la Cruz (Eds.), *Changing the faces of mathematics: Perspectives on Latinos* (pp. 71–80). Reston: National Council of Teachers of Mathematics.

Kilpatrick, J., Swafford, J., & Findell, B. (Eds.). (2001). *Adding it up: Helping children to learn mathematics*. Washington, DC: National Academy Press.

King, A. (1991). Effects of training in strategic questioning on children's problem-solving performance. *Journal of Educational Psychology, 83*(3), 307–317.

King, T. (2003). *The truth about stories: A native narrative*. Toronto: House of Anansi Press.

Kitchen, R. (2003). Getting real about mathematics education reform in high-poverty communities. *For the Learning of Mathematics, 23*(3), 16–22.

Knijnik, G. (2002). Ethnomathematics: Culture and politics of knowledge in mathematics education. *For the Learning of Mathematics, 22*(1), 11–14.

Knockwood, I. (1992). *Out of the depths: The experiences of Mi'kmaw children at the Indian Residential School at Shubenacadie, Nova Scotia* (2nd ed.). Lockeport: Roseway.

Kozulin, A., Gindis, B., Ageyev, V., & Miller, S. (Eds.). (2003). *Vygotsky's educational theory in cultural context*. Cambridge: Cambridge University Press.

Kress, G., & van Leeuwen, T. (2001). *Multimodal discourse: The modes and media of contemporary communication*. London: Edward Arnold.

Ladson-Billings, G. (1994). *The dreamkeepers: Successful teachers of African-American children*. San Francisco: Jossey-Bass Publishers.

Ladson-Billings, G. (1995a). Making mathematics meaningful in multicultural contexts. In W. Secada, E. Fennema, & L. Adajian (Eds.), *New directions for equity in mathematics education* (pp. 126–145). New York: Cambridge University Press.

Ladson-Billings, G. (1995b). But that's just good teaching!: The case for culturally relevant pedagogy. *Theory into Practice, 34*(3), 159–165.

Laghribi, A. (2005). Involutions en degré au plus 4 et corps des fonctions d'une quadrique en caractéristique 2. *Bulletin of the Belgian Mathematical Society – Simon Stevin, 12*(2), 161–174.

Lampert, M. (1990). When the problem is not the question and the solution is not the answer: Mathematical knowing and teaching. *American Educational Research Journal, 27*(1), 29–63.

Lampert, M. (1992). Practices and problems in teaching authentic mathematics in school. In F. Oser, A. Dick, & J.-L. Patry (Eds.), *Effective and responsible teaching: The new synthesis* (pp. 295–314). San Francisco: Jossey-Bass Publishers.

Lampert, M. (2001). *Teaching problems and the problems of teaching*. New Haven: Yale University Press.

Lampert, M., & Cobb, P. (2003). Communication and language. In J. Kilpatrick, W. Martin, & D. Schifter (Eds.), *A Research Companion to principles and standards for school mathematics* (pp. 237–249). Reston: National Council of Teachers of Mathematics.

Lave, J. (1988). *Cognition in practice: Mind, mathematics and culture in everyday life*. Cambridge: Cambridge University Press.

Lave, J. (1991). Situating learning in communities of practice. In L. Resnick, J. Levine, & S. Teasley (Eds.), *Perspectives on socially shared cognition* (pp. 63–82). Washington, DC: American Psychological Association.

Lave, J., & Wenger, E. (1991). *Situated learning: Legitimate peripheral participation*. Cambridge: Cambridge University Press.

Lave, J., Murtaugh, M., & de la Rocha, O. (1984). The dialectic of arithmetic in grocery shopping. In B. Rogoff & J. Lave (Eds.), *Everyday cognition: Its development in social context* (pp. 67–94). Cambridge: Harvard University Press.

Layzer, D. (1989). The synergy between writing and mathematics. In P. Connolly & T. Vilardi (Eds.), *Writing to learn mathematics and science* (pp. 122–133). New York: Teachers College Press.

Lee, C. (1993). *Signifying as a scaffold for literary interpretation: The pedagogical implications of an African American discourse genre*. Urbana: National Council of Teachers of English.

Lee, C. (2009). Historical evolution of risk and equity: Interdisciplinary issues and critiques. *Review of Research in Education, 33*(1), 63–100.

Lee, J. (2002). Racial and ethnic achievement gap trends: Reversing the progress towards equity? *Educational Researcher, 31*(1), 3–12.

Lee, J. (2004). Multiple facets of inequity in racial and ethnic achievement gaps. *Peabody Journal of Education, 79*(2), 51–73.

Lee, O., Deaktor, R., Hart, J., Cuevas, P., & Enders, C. (2005). An instructional intervention's impact on the science and literacy achievement of culturally and linguistically diverse elementary students. *Journal of Research in Science Teaching, 42*(8), 857–887.

Lemke, J. (1990). *Talking science: Language, learning, and values*. Norwood: Ablex Publishing Corporation.

Lerman, S. (2000). The social turn in mathematics education research. In J. Boaler (Ed.), *Multiple perspectives on mathematics teaching and learning* (pp. 19–44). Westport: Ablex Publishing Corporation.

Lerman, S. (2001). Cultural, discursive psychology: A sociocultural approach to studying the teaching and learning of mathematics. *Educational Studies in Mathematics, 46*(1–3), 87–113.

Lerman, S., & Tsatsaroni, A. (1998). Why children fail and what the field of mathematics education can do about it: The role of sociology. In P. Gates (Ed.), *Proceedings of the first International Mathematics Education and Society Conference* (pp. 26–33). Nottingham: Nottingham University, Centre for the Study of Mathematics Education.

Lerman, S., Tsatsaroni, A., Evans, J., Morgan, C., Kanes, C., Parker, D., & Adler, J. (2009). Sociological frameworks in mathematics education research: A research forum. In M. Tzekaki, M. Kaldrimidou, & H. Sakonidis (Eds.), *Proceedings of the 33rd Conference of the International Group for the Psychology of Mathematics Education* (Vol. 1, pp. 217–246). Thessaloniki: Aristotle University of Thessaloniki.

Leung, C. (2005). Language and content in bilingual education. *Linguistics and Education, 16*(2), 238–252.

Lewis, C. (2001). *Literary practices as social acts: Power, status, and cultural norms in the classroom*. Mahwah: Lawrence Erlbaum Associates.

Lipka, J., with Mohatt, G., & the Ciulistet group. (1998). *Transforming the culture of schools: Yup'ik Eskimo examples*. Mahwah: Lawrence Erlbaum Associates.

Lipman, P. (2004). *High stakes education: Inequality, globalization and urban school reform*. New York: RoutledgeFalmer.

Lobato, J., Clarke, D., & Ellis, A. (2005). Initiating and eliciting in teaching: A reformulation of telling. *Journal for Research in Mathematics Education, 36*(2), 101–136.

Lubienski, S. (2002). A closer look at black–white mathematics gaps: Intersections of race and SES in NAEP achievement and instructional practices data. *Journal of Negro Education, 71*(4), 269–287.

Lubienski, S. (2008). On "gap gazing" in mathematics education: The need for gaps analyses. *Journal for Research in Mathematics Education, 39*(4), 350–356.

Lubienski, S., & Gutiérrez, R. (2008). Bridging the gaps in perspectives on equity in mathematics education. *Journal for Research in Mathematics Education, 39*(4), 365–371.

Lubienski, S., McGraw, R., & Strutchens, M. (2004). NAEP findings regarding gender: Mathematics achievement, student affect, and learning practices. In P. Kloosterman & F. Lester (Eds.), *Results and interpretations of the 1990 through 2000 mathematics assessment of the National Assessment of Educational Progress* (pp. 305–336). Reston: National Council of Teachers of Mathematics.

Lunney Borden, L. (2010). *Transforming mathematics education for Mi'kmaw students through Mawikinutimatimk*. Unpublished Ph.D. dissertation, University of New Brunswick, Fredericton.

Lunney Borden, L. (2011). The 'verbification' of mathematics using the grammatical structures of Mi'kmaq to support student learning. *For the Learning of Mathematics, 31*(3), 2–7.

Lyddon Hatten, L. (2009). Talking around graphing calculators: A journey through performance gaps. In B. Herbel-Eisenmann & M. Cirillo (Eds.), *Promoting purposeful discourse: Teacher research in mathematics classrooms* (pp. 71–90). Reston: National Council of Teachers of Mathematics.

MacLure, M. (2003). *Discourse in educational and social research*. Buckingham: Open University Press.

MacSwan, J. (2000). The threshold hypothesis, semilingualism, and other contributions to a deficit view of linguistic minorities. *Hispanic Journal of Behavioral Sciences, 22*(91), 3–45.

Madaus, G. (1994). A technological and historical consideration of equity issues associated with proposals to change the nation's testing policy. *Harvard Educational Review, 64*(1), 76–95.

MALATI. (2005). *Mathematics learning and teaching initiative*. (www.academic.sun.ac.za/mathed/MALATI)

Males, L., Otten, S., & Herbel-Eisenmann, B. (2010). Challenges of critical colleagueship: Examining and reflecting on mathematics teacher study group interactions. *Journal of Mathematics Teacher Education, 13*(6), 459–471.

Malloy, C., & Malloy, W. (1998). Issues of culture in mathematics teaching and learning. *The Urban Review, 30*(3), 245–257.

Marks, J. (2009). Students' ownership and relinquishing control. In B. Herbel-Eisenmann & M. Cirillo (Eds.), *Promoting purposeful discourse: Teacher research in mathematics classrooms* (pp. 99–116). Reston: National Council of Teachers of Mathematics.

Martin, D. (2000). *Mathematics success and failure among African-American youth: The roles of sociohistorical context, community forces, school influence, and individual agency*. Mahwah: Lawrence Erlbaum Associates.

Martin, D. (2007). Mathematics learning and participation in the African American context: The co-construction of identity in two intersecting realms of experience. In N. Nasir & P. Cobb (Eds.), *Improving access to mathematics: Diversity and equity in the classroom* (pp. 146–158). New York: Teachers College Press.

Martin, D. (2009). Researching race in mathematics education. *Teachers College Record, 111*(2), 295–338.

Martin, D. (2010). Not-so-strange bedfellows: Racial projects and the mathematics education enterprise. In U. Gellert, E. Jablonka, & C. Morgan (Eds.), *Proceedings of the sixth international mathematics education and society conference* (pp. 57–79). Berlin: Freie Universität Berlin.

Martin, J., & Rose, D. (2003). *Working with discourse: Meaning beyond the clause*. London: Continuum.

Martin, J., & White, P. (2005). *The language of evaluation: Appraisal in English*. New York: Palgrave Macmillan.

McClain, K., & Cobb, P. (1997). An analysis of the teacher's role in guiding the evolution of sociomathematical norms. In E. Pekhonen (Ed.), *Proceedings of the 21st Conference of the International Group for the Psychology Mathematics Education* (Vol. 3, pp. 224–231). Lahti: University of Helsinki.

McDermott, R., & Varenne, H. (1995). Culture *as* disability. *Anthropology and Education Quarterly, 26*(3), 324–348.

McGaw, B. (2004). Australian mathematics learning in an international context. In I. Putt, R. Faragher, & M. McLean (Eds.), *Mathematics education for the third millennium: Towards 2010 (Proceedings of the 27th annual MERGA conference)* (pp. 639–663). Sydney: Mathematics Education Research Group of Australasia.

McGee Banks, C., & Banks, J. (1995). Equity pedagogy: An essential component of multicultural education. *Theory into Practice, 34*(3), 152–158.

McNeil, L., & Valenzuela, A. (2000). *The harmful impact of the TAAS system of testing in Texas: Beneath the accountability rhetoric*. Cambridge: Harvard Civil Rights Project.

Meaney, T., Trinick, T., & Fairhall, U. (2012). *Collaborating to meet language challenges in Indigenous mathematics classrooms*. New York: Springer.

Mehan, H. (1979). *Learning lessons*. Cambridge: Harvard University Press.

Mesa, V., & Chang, P. (2010). The language of engagement in two highly interactive undergraduate mathematics classrooms. *Linguistics and Education, 21*(2), 83–100.

Miller, J., Hoffer, T., Sucher, R., Brown, K., & Nelson, C. (1992). *LSAY code-book: Student, parent, and teacher data for 1992 cohort Two for longitudinal years one through four (1987–1991)*. DeKalb: Northern Illinois University.

Milloy, J. (1999). *A national crime: The Canadian government and the residential school system, 1879 to 1986*. Winnipeg: University of Manitoba Press.

Moll, L., Amanti, C., Neff, D., & González, N. (1992). Funds of knowledge for teaching: Using a qualitative approach to connect homes and classrooms. *Theory into Practice, 31*(2), 132–141.

Morales, H. (2007). *The underlife of a high school mathematics classroom: Mathematical meaning-making among Latino/a students*. Paper presented at the annual meeting of the American Educational Research Association, Chicago.

Morgan, C. (1998). *Writing mathematically: The discourse of investigation*. London: Falmer Press.

Morgan, C. (2005). Word, definitions and concepts in discourses of mathematics, teaching and learning. *Language and Education, 19*(2), 102–116.
Morgan, C. (2006). What does social semiotics have to offer mathematics education research? *Educational Studies in Mathematics, 61*(1–2), 219–245.
Morgan, C. (2007). Who is not multilingual now? *Educational Studies in Mathematics, 64*(2), 239–242.
Morgan, C., & Watson, A. (2002). The interpretative nature of teachers' assessment of students' mathematics: issues for equity. *Journal for Research in Mathematics Education, 33*(2), 78–110.
Morrow, W. (1993). Epistemological access in the university. *AD Issues, 1*(1), 3–4.
Moschkovich, J. (1996). Learning math in two languages. In L. Puig & A. Gutiérrez (Eds.), *Twentieth conference of the international group for the psychology of mathematics education* (Vol. 4, pp. 27–34). València: Universitat de València.
Moschkovich, J. (1999). Supporting the participation of English language learners in mathematical discussions. *For the Learning of Mathematics, 19*(1), 11–19.
Moschkovich, J. (2002). A situated and sociocultural perspective on bilingual mathematics learners. *Mathematical Thinking and Learning, 4*(2/3), 189–212.
Moschkovich, J. (2004). Appropriating mathematical practices: A case study of learning to use and explore functions through interaction with a tutor. *Educational Studies in Mathematics, 55*(1–3), 49–80.
Moschkovich, J. (2007a). Bilingual mathematics learners: How views of language, bilingual learners, and mathematical communication affect instruction. In N. Nasir & P. Cobb (Eds.), *Improving access to mathematics: Diversity and equity in the classroom* (pp. 89–104). New York: Teachers College Press.
Moschkovich, J. (2007b). Examining mathematical discourse practices. *For the Learning of Mathematics, 27*(1), 24–30.
Moschkovich, J. (2007c). Using two languages when learning mathematics. *Educational Studies in Mathematics, 64*(2), 121–144.
Moschkovich, J. (2008). "I went by twos, he went by one": Multiple interpretations of inscriptions as resources for mathematical discussions. *The Journal of the Learning Sciences, 17*(4), 551–587.
Moschkovich, J. (2010). Language(s) and learning mathematics: Resources, challenges, and issues for research. In J. Moschkovich (Ed.), *Language and mathematics education: Multiple perspectives and directions for research* (pp. 1–28). Charlotte: Information Age Publishing.
Moschkovich, J., & Brenner, M. (2002). Preface. In M. Brenner & J. Moschkovich (Eds.), *Everyday and academic mathematics in the classroom (JRME monograph number 11)* (pp. v–x). Reston: National Council of Teachers of Mathematics.
Moschkovich, J., & Nelson-Barber, S. (2009). What mathematics teachers need to know about culture and language. In B. Greer, S. Mukhopadhyay, A. Powell, & S. Nelson-Barber (Eds.), *Culturally responsive mathematics education* (pp. 111–136). New York: Routledge.
Mousley, J., & Marks, G. (1991). *Discourses in mathematics*. Geelong: Deakin University.
Mukhopadhyay, S., & Greer, B. (2001). Modeling with purpose: Mathematics as a critical tool. In B. Atweh, H. Forgasz, & B. Nebres (Eds.), *Sociocultural research on mathematics education: An international perspective* (pp. 295–311). Mahwah: Lawrence Erlbaum Associates.
Nasir, N., & Cobb, P. (Eds.). (2007). *Improving access to mathematics: Diversity and equity in the classroom.* New York: Teachers College Press.
Nasir, N., & Saxe, G. (2003). Ethnic and academic identities: A cultural practice perspective on emerging tensions and their management in the lives of minority students. *Educational Researcher, 32*(5), 14–18.
Nasir, N., Rosebery, A., Warren, B., & Lee, C. (2006). Learning as a cultural process: Achieving equity through diversity. In R. Sawyer (Ed.), *The Cambridge handbook of the learning sciences* (pp. 489–504). Cambridge: Cambridge University Press.
Nathan, M., & Knuth, E. (2003). A study of whole classroom mathematical discourse and teacher change. *Cognition and Instruction, 21*(2), 175–207.

Nattiv, A. (1994). Helping behaviors and math achievement gain of students using cooperative learning. *The Elementary School Journal, 94*(3), 285–297.
NBDOE. (2001). *Atlantic Canada mathematics curriculum: Mathematics 5*. Fredericton: New Brunswick Department of Education.
NCTM. (1989). *Curriculum and evaluation standards*. Reston: National Council of Teachers of Mathematics.
NCTM. (1991). *Professional standards for teaching mathematics*. Reston: National Council of Teachers of Mathematics.
NCTM. (2000). *Principles and standards for school mathematics*. Reston: National Council of Teachers of Mathematics.
Nelson-Barber, S., & Lipka, J. (2008). Rethinking the case for culture-based curriculum: Conditions that support improved mathematics performance in diverse classrooms. In M. Brisk (Ed.), *Language, culture, and community in teacher education* (pp. 99–123). Mahwah: Lawrence Erlbaum Associates.
Nemirovsky, R., DiMattia, C., Ribeiro, B., & Lara-Meloy, T. (2005). Talking about teaching episodes. *Journal of Mathematics Teacher Education, 8*(5), 363–392.
NRC. (2004). *Engaging schools: Fostering high school students' motivation to learn*. Washington, DC: National Academies Press.
NSDOE. (2002). *Foundation for Mi'kmaw/Miigmao language curriculum*. Halifax: Nova Scotia Department of Education.
Nunes, T., Schliemann, A., & Carraher, D. (1993). *Street mathematics and school mathematics*. Cambridge: Cambridge University Press.
Nystrand, M. (1997). *Opening dialogue: Understanding the dynamics of language and learning in the English classroom*. New York: Teachers College Press.
O'Connor, C., Hill, L., & Robinson, S. (2009). Who's at risk in school and what's race got to do with it? *Review of Research in Education, 33*(1), 1–34.
O'Connor, M. (1998). Language socialization in the mathematics classroom: Discourse practices and mathematical thinking. In M. Lampert & M. Blunk (Eds.), *Talking mathematics in school: Studies of teaching and learning* (pp. 17–55). New York: Cambridge University Press.
O'Connor, M., & Michaels, S. (1993). Aligning academic task and participation status through revoicing: Analysis of a classroom discourse strategy. *Anthropology and Education Quarterly, 24*(4), 318–335.
O'Connor, M., & Michaels, S. (1996). Shifting participant frameworks: Orchestrating thinking practices in group discussion. In D. Hicks (Ed.), *Discourse, learning and schooling* (pp. 63–103). New York: Cambridge University Press.
O'Connor, M., Godfrey, L., & Moses, R. (1998). The missing data point: Negotiating purposes in classroom mathematics and science. In J. Greeno & S. Goldman (Eds.), *Thinking practices in mathematics and science learning* (pp. 89–125). Mahwah: Lawrence Erlbaum Associates.
O'Halloran, K. (1998). Classroom discourse in mathematics: A multisemiotic analysis. *Linguistics and Education, 10*(3), 359–388.
O'Halloran, K. (2004). Discourses in secondary school mathematics classrooms according to social class and gender. In J. Foley (Ed.), *Language, education and discourse: Functional approaches* (pp. 191–225). London: Continuum.
O'Halloran, K. (2005). *Mathematical discourse: Language, symbolism and visual images*. New York: Continuum.
Oakes, J. (1995). Two cities' tracking and within-school segregation. *Teachers College Record, 96*(4), 681–690.
Oakes, J., Gamoran, A., & Page, R. (1992). Curriculum differentiation: Opportunities, outcome and meanings. In P. Jackson (Ed.), *Handbook of research on curriculum* (pp. 570–608). New York: Macmillan.
Oakes, J., Muir, M., & Joseph, R. (2000). *Course-taking and achievement in mathematics and science: Inequalities that endure and change*. Madison: National Institute of Science Education.

Obrycki, J. (2009). Listening to my students' thoughts on mathematics education. In B. Herbel-Eisenmann & M. Cirillo (Eds.), *Promoting purposeful discourse: Teacher research in mathematics classrooms* (pp. 187–202). Reston: National Council of Teachers of Mathematics.

Orr, E. (1987). *Twice as less: Black English and the performance of black students in mathematics and science*. New York: W. W. Norton.

Orr, J., Paul, J., & Paul, S. (2002). Decolonizing Mi'kmaw education through cultural practical knowledge. *McGill Journal of Education, 37*(3), 331–354.

Paine, L., Fang, Y., Wilson, S., et al. (2003). Entering a culture of teaching: Teacher induction in shanghai. In E. Britton (Ed.), *Comprehensive teacher induction: Systems for early career learning* (pp. 20–82). Dordrecht: Kluwer Academic Publishers.

Perry, W. (1970/1999). *Forms of intellectual and ethical development in the college years: A scheme*. San Francisco: Jossey-Bass Publishers.

Pimm, D. (1987). *Speaking mathematically: Communication in mathematics classrooms*. London: Routledge and Kegan Paul.

Pimm, D. (1994). Spoken mathematics classroom culture: Artifice and artificiality. In S. Lerman (Ed.), *Cultural perspectives on the mathematics classroom* (pp. 133–147). Dordrecht: Kluwer Academic Publishers.

Pimm, D. (1995). *Symbols and meanings in school mathematics*. London: Routledge.

Pimm, D. (2007). Registering surprise. *For the Learning of Mathematics, 27*(1), 31.

Pimm, D. (2003). Being and becoming a mathematics teacher: Ambiguities in teacher formation in France. In E. Britton et al. (Eds.), *Comprehensive teacher induction: Systems for early career learning* (pp. 194–260). Dordrecht: Kluwer Academic Publishers.

Pirandello, L. (1922/1931). Six characters in search of an author. In *Three plays*. New York: E. P. Dutton.

PNB. (1997). *Education act*. Fredericton: Province of New Brunswick. (http://www.gnb.ca/0062/PDF-acts/e-01-12.pdf)

Powell, A., & Frankenstein, M. (Eds.). (1997). *Ethnomathematics: Challenging Eurocentrism in mathematics education*. Albany: State University of New York Press.

Pressley, M. (2000). What should comprehension instruction be the instruction of? In M. Kamil, P. Mosenthal, P. Pearson, & R. Barr (Eds.), *Handbook of Reading Research* (Vol. 3, pp. 545–561). Mahwah: Lawrence Erlbaum Associates.

Prince, E., Frader, J., & Bosk, C. (1982). On hedging in physician–physician discourse. In R. DiPietro (Ed.), *Linguistics and the professions* (pp. 83–97). Norwood: Ablex Publishing Corporation.

Pullin, D., & Haertel, E. (2008). Assessment through the lens of "opportunity to learn". In P. Moss, D. Pullin, J. Gee, E. Haertel, & L. Young (Eds.), *Assessment, equity and opportunity to learn* (pp. 17–41). New York: Cambridge University Press.

Radford, L., Bardini, C., & Sabena, C. (2007). Perceiving the general: The multisemiotic dimension of students' algebraic activity. *Journal for Research in Mathematics Education, 38*(5), 507–530.

Reed, R., & Oppong, N. (2005). Looking critically at teachers' attention to equity in their classrooms. *The Mathematics Educator Monograph no. 1* (pp. 2–15). (math.coe.uga.edu/tme/issues/monograph1/mono1_reedabs.html)

Roberts, T. (1998). Mathematical registers of aboriginal languages. *For the Learning of Mathematics, 18*(1), 10–16.

Rotman, B. (2000). *Mathematics as sign: Writing, imagining, counting*. Palo Alto: Stanford University Press.

Rousseau, C., & Powell, A. (2005). Understanding the significance of context: A framework to examine equity and reform in secondary mathematics. *The High School Journal, 88*(4), 19–31.

Rowland, T. (1995). Hedges in mathematics talk: Linguistic pointers to uncertainty. *Educational Studies in Mathematics, 29*(4), 327–353.

Rowland, T. (1999). Pronouns in mathematics talk: Power, vagueness and generalisation. *For the Learning of Mathematics, 19*(2), 19–26.

Rowland, T. (2000). *The pragmatics of mathematics education: Vagueness in mathematical discourse*. London: Falmer Press.

Ryve, A. (2011). Discourse research in mathematics education: A critical evaluation of 108 journal articles. *Journal for Research in Mathematics Education, 42*(2), 167–198.

Sachs, A. (1994). *Language rights in the new constitution*. Belleville: University of the Western Cape, South Africa Constitution Studies Centre.

Sacks, H. (1992). *Lectures on conversation*. Oxford: Blackwell.

Schleppegrell, M. (2004). *The language of schooling: A functional linguistics perspective*. Mahwah: Lawrence Erlbaum Associates.

Schleppegrell, M. (2010). Language in mathematics teaching and learning: A research review. In J. Moschkovich (Ed.), *Language and mathematics education: Multiple perspectives and directions for research* (pp. 73–112). Charlotte: Information Age Publishing.

Scribner, S. (1984). Studying working intelligence. In B. Rogoff & J. Lave (Eds.), *Everyday cognition: Its development in social context* (pp. 9–40). Cambridge: Harvard University Press.

Secada, W. (1989). Educational equity versus equality of education: An alternative conception. In W. Secada (Ed.), *Equity in education* (pp. 68–88). Philadelphia: Falmer Press.

Secada, W., Hankes, J., & Fast, G. (Eds.). (2002). *Changing the faces of mathematics: Perspectives of indigenous people of North America*. Reston: National Council of Teachers of Mathematics.

Seeger, F., Voigt, J., & Waschescio, U. (Eds.). (1998). *The culture of the mathematics classroom*. Cambridge: Cambridge University Press.

Setati, M. (1998). Code-switching in a senior primary class of second-language mathematics learners. *For the Learning of Mathematics, 18*(1), 34–40.

Setati, M. (2005a). Teaching mathematics in a primary multilingual classroom. *Journal for Research in Mathematics Education, 36*(5), 447–466.

Setati, M. (2005b). Power and access in multilingual mathematics classrooms. In M. Goos, C. Kanes, & R. Brown (Eds.), *Proceedings of the Fourth International Mathematics Education and Society Conference* (pp. 7–18). Brisbane: Griffith University.

Setati, M. (2008). Access to mathematics versus access to the language of power: The struggle in multilingual classrooms. *South African Journal of Education, 28*(1), 103–116.

Setati, M., & Adler, J. (2001). Between languages and discourses: Code-switching practices in primary multilingual mathematics classrooms in South Africa. *Educational Studies in Mathematics, 43*(3), 243–269.

Setati, M., Chitera, N., & Essien, A. (2009). Research on multilingualism in mathematics education in South Africa: 2000–2007. *African Journal of Research in Mathematics, Science and Technology Education, 13*(1), 64–79.

Setati, M., Molefe, T., & Langa, M. (2008). Using language as a transparent resource in the teaching and learning of mathematics in a grade 11 multilingual classroom. *Pythagoras, 67*, 14–25.

Sfard, A. (2008). *Thinking as communicating: Human development, the growth of discourses, and mathematizing*. Cambridge: Cambridge University Press.

Sfard, A., Forman, E., & Kieran, C. (Eds.) (2001). Bridging the individual and the social: Discursive approaches to research in mathematics education. *Educational Studies in Mathematics 46*(1–3), 1–306.

Sfard, A., Nesher, P., Streefland, L., Cobb, P., & Mason, J. (1998). Learning mathematics through conversation: Is it as good as they say? *For the Learning of Mathematics, 18*(1), 41–51.

Shepard, L. (1991). Will national tests improve student learning? *Phi Delta Kappan, 73*(3), 232–238.

Sherin, M., & Han, S. (2004). Teacher learning in the context of a video club. *Teaching and Teacher Education, 20*(2), 163–183.

Shreyar, S., Zolkower, B., & Pérez, S. (2010). Thinking aloud together: A teacher's semiotic mediation of a whole-class conversation about percents. *Educational Studies in Mathematics, 73*(1), 21–53.

SIA. (2003). Why don't kids learn maths and science successfully?. *Science in Africa, 27*. (www.scienceinafrica.co.za/june/maths.htm)

Siemon, D. (2008). *Developing the "Big" ideas in mathematics*. (http://www.decs.sa.gov.au/learninginclusion/files/pages/Expo/Siemen/Devt_Big_Ideas_in_Mathemat.pdf)

Silverman, D. (1998). *Harvey Sacks: Social science and conversation analysis*. New York: Oxford University Press.

Sinclair, J., & Coulthard, M. (1975). *Towards an analysis of discourse: The English used by teachers and pupils*. Oxford: Oxford University Press.

Sinclair, N., & Pimm, D. (2010). The many and the few: Mathematics, democracy and the aesthetic. *Educational Insights 13*(1). (www.ccfi.educ.ubc.ca/publication/insights/)

Singh, P. (2002). Pedagogising knowledge: Bernstein's theory of the pedagogic device. *British Journal of Sociology of Education, 23*(4), 571–582.

Siskin, L., & Little, J. (Eds.). (1995). *The subjects in question: Departmental organization and the high school*. New York: Teachers College Press.

Skovsmose, O. (1994). *Towards a philosophy of critical mathematics education*. Dordrecht: Kluwer Academic Publishers.

Skovsmose, O., & Borba, M. (2004). Research methodology and critical mathematics education. In P. Valero & R. Zevenbergen (Eds.), *Researching the socio-political dimensions of mathematics education: Issues of power in theory and methodology* (pp. 207–226). Dordrecht: Kluwer Academic Publishers.

Skovsmose, O., & Valero, P. (2001). Breaking political neutrality: The critical engagement of mathematics education with democracy. In B. Atweh, H. Forgasz, & B. Nebres (Eds.), *Sociocultural research on mathematics education: An international perspective* (pp. 37–55). Mahwah: Lawrence Erlbaum Associates.

Smith, J. (1996). Efficacy and teaching mathematics by telling: A challenge for reform. *Journal for Research in Mathematics Education, 27*(4), 387–402.

Spears, A. (2007). African American communicative practices: Improvisation, semantic license, and augmentation. In H. Alim & J. Baugh (Eds.), *Talkin Black Talk: Language, education, and social change* (pp. 100–111). New York: Teachers College Press.

Spillane, J. (1998). State policy and the non-monolithic nature of the local school district: Organizational and professional considerations. *American Educational Research Journal, 35*(1), 33–63.

Spindler, G., & Spindler, L. (1997). Ethnography: An anthropological view. In G. Spindler (Ed.), *Education and cultural process: Anthropological approaches* (pp. 151–156). Prospect Heights: Waveland Press.

Staples, M., & Truxaw, M. (2010). The *Mathematics Learning Discourse* project: Fostering higher order thinking and academic language in urban mathematics classrooms. *Journal of Urban Mathematics Education, 3*(1), 27–56.

Stein, M., Silver, E., & Smith, M. (1998). Mathematics reform and teacher development: A community of practice perspective. In J. Greeno & S. Goldman (Eds.), *Thinking practices in mathematics and science learning* (pp. 17–52). Mahwah: Lawrence Erlbaum Associates.

Stein, M., Smith, M., Henningsen, M., & Silver, E. (2000). *Implementing standards-based mathematics instruction: A casebook for professional development*. Reston: National Council of Teachers of Mathematics.

Steinbring, H., Bartolini Bussi, M., & Sierpinska, A. (Eds.). (1998). *Language and communication in the mathematics classroom*. Reston: National Council of Teachers of Mathematics.

Stevens, F. (1993). Applying an opportunity-to-learn conceptual framework to the investigation of the effects of teaching practices via secondary analyses of multiple-case-study summary data. *The Journal of Negro Education, 62*(3), 232–248.

Stevens, R., & Slavin, R. (1995). The cooperative elementary school: Effects on students' achievement, attitudes, and social relations. *American Educational Research Journal, 32*(2), 321–351.

Stigler, J., & Hiebert, J. (1999). *The teaching gap: Best ideas from the world's teachers for improving education in the classroom*. New York: The Free Press.

Stodolsky, S., & Grossman, P. (1995). The impact of subject matter on curricular activity: An analysis of five academic subjects. *American Educational Research Journal, 32*(2), 227–249.

Swain, M. (2000). The output hypothesis and beyond: Mediating acquisition through collaborative dialogue. In J. Lantolf (Ed.), *Sociocultural theory and second language learning* (pp. 97–114). Oxford: Oxford University Press.

Tahta, D. (1991). Understanding and desire. In D. Pimm & E. Love (Eds.), *Teaching and learning school mathematics* (pp. 221–246). London: Hodder and Stoughton.

Talbert, J. (1995). Boundaries of teachers' professional communities in U.S. high schools: Power and precariousness of the subject department. In L. Siskin & J. Little (Eds.), *The subjects in question: Departmental organization and the high school* (pp. 68–94). New York: Teachers College Press.

Taylor, N., & Vinjevold, P. (Eds.) (1999). *Getting learning right* (Report of the President's Education Initiative Research Project). Johannesburg: Joint Education Trust.

ten Have, P. (1999). *Doing conversation analysis: A practical guide*. London: Sage.

Thomas, W., & Collier, V. (1997). *School effectiveness for language minority students*. Washington, DC: National Clearinghouse for Bilingual Education.

Tompkins, J. (2002). Learning to see what they can't: Decolonizing perspectives on indigenous education in the racial context of rural Nova Scotia. *McGill Journal of Education, 37*(3), 405–422.

Tyack, D., & Cuban, L. (1995). *Tinkering toward utopia: A century of public school reform*. Cambridge: Harvard University Press.

USDOE. (1998). *Latinos in education: Early childhood, elementary, secondary, undergraduate, graduate* (White House initiative on educational excellence for Hispanic Americans). Washington, DC: U.S. Department of Education. (http://permanent.access.gpo.gov/lps15101/fact-1.pdf)

Valdés-Fallis, G. (1978). Code switching and the classroom teacher. In *Language in education: Theory and practice* (Vol. 4, pp. 1–26). Arlington: Center for Applied Linguistics (ERIC Document Reproduction Service No. ED153506).

Valdés-Fallis, G. (1979). Social interaction and code switching patterns: A case study of Spanish/English alternation. In G. Keller, R. Teichner, & S. Viera (Eds.), *Bilingualism in the bicentennial and beyond* (pp. 86–96). Jamaica: Bilingualism Press.

Valero, P., & Zevenbergen, R. (Eds.). (2004). *Researching the socio-political dimensions of mathematics education: Issues of power in theory and methodology*. Dordrecht: Kluwer Academic Publishers.

van Langenhove, L., & Harré, R. (1999). Positioning as the production and use of stereotypes. In R. Harré & L. van Langenhove (Eds.), *Positioning theory: Moral contexts of intentional action* (pp. 127–137). Oxford: Blackwell.

Varghese, T. (2009). Teaching mathematics with a holistic approach. *International Journal of Inclusive Education, 13*(1), 13–22.

Vithal, R., & Skovsmose, O. (1997). The end of innocence: A critique of ethnomathematics. *Educational Studies in Mathematics, 34*(2), 131–157.

Vogt, L., Jordan, C., & Tharp, R. (1987). Explaining school failure, producing school success: Two cases. *Anthropology and Education Quarterly, 18*(4), 276–286.

Voigt, J. (1985). Patterns and routines in classroom interaction. *Recherches en Didactique des Mathématiques, 6*(1), 69–118.

Voigt, J. (1989). Social functions of routines and consequences for subject matter learning. *International Journal of Educational Research, 13*(6), 647–656.

Vygotsky, L. (1978). *Mind in society: The development of higher psychological processes*. Cambridge: Harvard University Press.

Vygotsky, L. (1986, rev'd edn). *Thought and language*. Cambridge: Massachusetts Institute of Technology Press.

Wagner, D. (2007). Students' critical awareness of voice and agency in mathematics classroom discourse. *Mathematical Thinking and Learning, 9*(1), 31–50.

Wagner, D. (2008). "Just go": Mathematics students' critical awareness of routine procedure. *Canadian Journal of Science, Mathematics and Technology Education, 8*(1), 35–48.

Wagner, D., & Herbel-Eisenmann, B. (2008). "Just don't": The suppression and invitation of dialogue in mathematics classrooms. *Educational Studies in Mathematics, 67*(2), 143–157.

Wagner, D., & Herbel-Eisenmann, B. (2009). Re-mythologizing mathematics through attention to classroom positioning. *Educational Studies in Mathematics, 72*(1), 1–15.
Wagner, D., & Lunney, L. (2006). Common sense, necessity, and intention in ethnomathematics. In S. Alatorre, J. Cortina, M. Sáiz, & Méndez, A. (Eds.), *Proceedings of the 28th annual meeting of the North American Chapter of the International Group for the Psychology of Mathematics Education* (vol. II, 521-523), Mérida: Universidad Pedagógica Nacional.
Wagner, D., & Lunney Borden, L. (2011). Qualities of respectful positioning and their connections to quality mathematics. In B. Atweh, M. Graven, W. Secada, & P. Valero (Eds.), *Mapping equity and quality in mathematics education* (pp. 379–391). New York: Springer.
Wagner, D., & Lunney Borden, L. (in press). Common sense and necessity in (ethno)mathematics. In K. Sullenger & S. Turner (Eds.), *CRYSTAL Atlantique: A story of informal learning research in science, mathematics, and technology*. Rotterdam: Sense Publishers.
Walkerdine, V. (1988). *The mastery of reason: Cognitive development and the production of rationality*. London: Routledge.
Walkerdine, V., & Lucey, H. (1989). *Democracy in the kitchen: Regulating mothers and socialising daughters*. London: Virago Press.
Wang, J., & Paine, L. (2003). Learning to teach with mandated curriculum and public examination of teaching as contexts. *Teaching and Teacher Education, 19*(1), 75–94.
Warren, B., & Rosebery, A. (1995). Equity in the future tense: redefining relationships among teachers, students, and science in linguistic minority classrooms. In W. Secada, E. Fennema, & L. Adajian (Eds.), *New directions for equity in mathematics education* (pp. 298–328). New York: Cambridge University Press.
Watson, H. (1987). Learning to apply numbers to nature. *Educational Studies in Mathematics, 18*(4), 339–357.
Watson, H., & Chambers, D. (1989). *Singing the land, signing the land: A portfolio of exhibits*. Geelong: Deakin University.
Webb, N. (1991). Task-related verbal interaction and mathematics learning in small groups. *Journal for Research in Mathematics Education, 22*(5), 366–389.
Webb, N., & Mastergeorge, A. (2003). The development of students' helping behavior and learning in peer-directed small groups. *Cognition and Instruction, 21*(4), 361–428.
Weick, K. (1976). Educational organizations as loosely coupled systems. *Administrative Science Quarterly, 21*(1), 1–19.
Weingrad, P. (1998). Teaching and learning politeness for mathematical argument in school. In M. Lampert & M. Blunk (Eds.), *Talking mathematics in school: Studies of teaching and learning* (pp. 213–239). New York: Cambridge University Press.
Weir, A. (1974). *General integration and measure*. Cambridge: Cambridge University Press.
Weiss, I. (1994). *A profile of science and mathematics education in the United States, 1993*. Chapel Hill: Horizon Research, Inc. (http://2000survey.horizon-research.com/reports/profile93.pdf)
Weissglass, J. (2000). No compromise on equity in mathematics education: Developing an infrastructure. In W. Secada (Ed.), *Changing the faces of mathematics: Perspectives on multiculturalism and gender equity* (pp. 5–24). Reston: National Council of Teachers of Mathematics.
Wenger, E. (1998). *Communities of practice: Learning, meaning, and identity*. Cambridge: Cambridge University Press.
Wertsch, J. (1984). The zone of proximal development: Some conceptual issues. In B. Rogoff & J. Wertsch (Eds.), *Children's Learning in the "Zone of Proximal Development"* (pp. 7–18). San Francisco: Jossey-Bass Publishers.
Wertsch, J. (1985). *Vygotsky and the social formation of mind*. Cambridge: Harvard University Press.
Wetherell, M. (2007). A step too far: Discursive psychology, linguistic ethnography and questions of identity. *Journal of Sociolinguistics, 11*(5), 661–681.
Whitenack, J., & Yackel, E. (2002). Making mathematical arguments in the primary grades: The importance of explaining and justifying ideas. *Teaching Children Mathematics, 8*(9), 524–527.
Willis, S. (2000). Strengthening numeracy: Reducing risk. In M. Meiers (Ed.), *Proceedings of the improving numeracy learning research conference 2000*. Melbourne: Australian Council for Educational Research.

Wood, T. (1998). Alternative patterns of communication in mathematics classes: Funneling or focusing? In H. Steinbring, M. Bartolini Bussi, & A. Sierpinska (Eds.), *Language and communication in the mathematics classroom* (pp. 167–178). Reston: National Council of Teachers of Mathematics.

Wortham, S. (2004). From good student to outcast: The emergence of a classroom identity. *Ethos, 32*(2), 164–187.

Yackel, E., & Cobb, P. (1996). Sociomathematical norms, argumentation, and autonomy in mathematics. *Journal for Research in Mathematics Education, 27*(4), 458–477.

Zack, V., & Graves, B. (2001). Making mathematical meaning through dialogue: "Once you think of it, the Z minus three seems pretty weird". *Educational Studies in Mathematics, 46*(1–3), 229–271.

Zentella, A. (1997). *Growing up bilingual: Puerto Rican children in New York.* Malden: Blackwell.

Zevenbergen, R. (1995). *The construction of social difference in mathematics education.* Unpublished Ph.D. dissertation, Deakin University, Geelong.

Zevenbergen, R. (2000). "Cracking the code" of mathematics classrooms: School success as a function of linguistic, social, and cultural background. In J. Boaler (Ed.), *Multiple perspectives on mathematics teaching and learning* (pp. 201–223). Westport: Ablex Publishing Corporation.

Zevenbergen, R. (2001a). Mathematical literacy in the middle years. *Literacy in the Middle Years, 9*(2), 21–28.

Zevenbergen, R. (2001b). Mathematics, social class, and linguistic capital: An analysis of mathematics classroom interactions. In B. Atweh, H. Forgasz, & B. Nebres (Eds.), *Sociocultural research on mathematics education: An international perspective* (pp. 201–215). Mahwah: Lawrence Erlbaum Associates.

Zevenbergen, R. (2005). The construction of a mathematical habitus: Implications of ability grouping in the middle years. *Journal of Curriculum Studies, 37*(5), 607–619.

Zevenbergen, R., & Lerman, S. (2001). Communicative competence in school mathematics: On being able to *do* school mathematics. In J. Bobis, B. Perry, & M. Mitchelmore (Eds.), *Numeracy and beyond (Proceedings of the 24th Annual MERGA Conference of the Australasia)* (Vol. 2, pp. 571–578). Sydney: Mathematics Education Research Group of Australasia.

Zevenbergen, R., Mousley, J., & Sullivan, P. (2004). Making the pedagogic relay inclusive for indigenous Australian students in mathematics classrooms. *International Journal of Inclusive Education, 8*(4), 391–405.

Zolkower, B., & Shreyar, S. (2007). A teacher's mediation of a thinking-aloud discussion in a 6th grade mathematics classroom. *Educational Studies in Mathematics, 65*(2), 177–202.

Author Index

B. Herbel-Eisenmann et al. (eds.), *Equity in Discourse for Mathematics Education: Theories, Practices, and Policies*, Mathematics Education Library 55, DOI 10.1007/978-94-007-2813-4, © Springer Science+Business Media B.V. 2012

Subject Index

B. Herbel-Eisenmann et al. (eds.), *Equity in Discourse for Mathematics Education: Theories, Practices, and Policies*, Mathematics Education Library 55,
DOI 10.1007/978-94-007-2813-4, © Springer Science+Business Media B.V. 2012

V

W

MIX
Papier aus verantwortungsvollen Quellen
Paper from responsible sources
FSC® C105338
FSC
www.fsc.org

If you have any concerns about our products,
you can contact us on
ProductSafety@springernature.com

In case Publisher is established outside the EU,
the EU authorized representative is:
Springer Nature Customer Service Center GmbH
Europaplatz 3, 69115 Heidelberg, Germany

Printed by Libri Plureos GmbH
in Hamburg, Germany